COMPUTATIONAL THINKING
CURRICULA IN K–12

COMPUTATIONAL THINKING CURRICULA IN K–12

INTERNATIONAL IMPLEMENTATIONS

EDITED BY HAROLD ABELSON AND SIU-CHEUNG KONG

THE MIT PRESS   CAMBRIDGE, MASSACHUSETTS   LONDON, ENGLAND

The MIT Press would like to thank the anonymous peer reviewers who provided comments on drafts of this book. The generous work of academic experts is essential for establishing the authority and quality of our publications. We acknowledge with gratitude the contributions of these otherwise uncredited readers.

This book was set in Stone Serif by Westchester Publishing Services. Printed and bound in the United States of America.

Library of Congress Cataloging-in-Publication Data

Names: Abelson, Harold, editor. | Kong, Siu Cheung, editor.
Title: Computational thinking curricula in K-12 : international
   implementations / edited by Harold Abelson and Siu-Cheung Kong.
Description: Cambridge, Massachusetts : The MIT Press, [2024] |
   Includes bibliographical references and index.
Identifiers: LCCN 2023027971 (print) | LCCN 2023027972 (ebook) |
   ISBN 9780262548052 (paperback) | ISBN 9780262378659 (epub) |
   ISBN 9780262378642 (pdf)
Subjects: LCSH: Computer science—Study and teaching—Case studies. |
   Problem solving—Study and teaching—Case studies.
Classification: LCC QA76.27 .C478 2024 (print) | LCC QA76.27 (ebook) |
   DDC 004.071—dc23/eng/20230905
LC record available at https://lccn.loc.gov/2023027971
LC ebook record available at https://lccn.loc.gov/2023027972

ISBN: 978-0-262-54805-2

10 9 8 7 6 5 4 3 2 1

CONTENTS

INTRODUCTION

Siu-Cheung Kong, Harold Abelson, and Wai-Ying Kwok

Educational authorities around the world share the consensus that timely and flexible policies should be made for fostering students to start their journey of computational thinking education (CTE) in schooling life (Grover and Pea 2013; Hsu, Irie, and Ching 2019). This consensus leads to the growing need and trend of implementing computational thinking (CT) curricula in the K–12 sector around the world (Borrás and Edquist 2015; Hsu et al. 2019). The effective implementation of CT curricula in the K–12 school contexts requires coherent efforts in educational policy and curriculum development in CTE in K–12 as well as in school implementation and classroom practice of CT curricula in K–12 (Mao, Ifenthaler, Fujimoto, Garavaglia, and Rossi 2019; Williamson, Bergviken Rensfeldt, Player-Koro, and Selwyn 2019). All these efforts are important issues and deserve further deliberation in the K–12 sector. This book presents an overview of these important issues for CTE in K–12, with contributions from experts in twelve countries/regions across three continents.

This book emerged from our earlier work, *Computational Thinking Education in K–12: Artificial Intelligence Literacy and Physical Computing.* This previous collection explored classroom-level experiments and implementations of teaching CT. In the course of assembling the earlier book, we also received contributions that looked at national-level curricular implementations. These latter contributions, and the separate set of issues they explored, encouraged us to solicit chapters from a wider range of countries to present an early overview of CT education at a national policy level.

While more work remains to be done in providing a truly systematic and consistent analysis of CT curricula around the world, these essays provide insight into the issues, hurdles, and potentials of introducing comprehensive new curricular material in a coherent, developmentally appropriate progression across primary and secondary school grades. The chapters in this book are organized into two parts to provide policymakers, curriculum developers, school practitioners, and educational researchers with a global overview of the two focal areas important for the planning, development, and implementation of CT curricula in K–12.

PART I: EDUCATIONAL POLICY AND CURRICULUM DEVELOPMENT FOR COMPUTATIONAL THINKING EDUCATION IN K–12

The first part of this volume presents a global deliberation of the policy landscape and school planning of CT curricula in K–12. Six chapters contributed by leaders in CTE in Asia, Europe, and Oceania provide an extensive discussion in relation to this focus area.

In "Computational Thinking Education in the UK: A Teacher's Perspective," Scott provides a holistic overview and reflection on the policy and implementation of national CT curricula in K–12 schools in the United Kingdom. This chapter leads readers to start the volume-reading journey by looking into how the United Kingdom, a country with a long development history and many invention records in the field of computing, offers and evolves CTE among its four constituting nations—England, Scotland, Wales, and Northern Ireland. The author provides a systematic outline, conducts a meaningful comparison, and makes a critical reflection on the different approaches taken in the four nations for planning, developing, and implementing CT curricula in K–12.

The second chapter in this part—"Computational Thinking Education in Hong Kong" by Kong and Kwok—discusses and illustrates the planning, development, and implementation of the CT curriculum for K–12 schools in Hong Kong. The chapter introduces readers to a worldwide-first initiative in an important hub of e-learning in the Asia-Pacific region for pioneering the curriculum development for CTE at senior primary grades.

The authors holistically depict the policy landscape of CTE in school education in Hong Kong, elaborately illustrate the design and implementation of the citywide CT curriculum pilot, and critically reflect on the lessons learned through the Hong Kong–based CT curriculum pilot.

"A Core-Competency Oriented Computational Thinking Education in China" by Yang and Ren introduces and illustrates how educational authorities in China adopt a core-competency-oriented perspective for the curriculum development and implementation for CTE in K–12. This chapter guides readers to look into the governmental directions and strategic plans of CT curricula in K–12 in one of the fastest-growing economies in the world in recent decades. The authors deliberate the history, rationale, and delivery of national initiatives related to curricula for CTE in K–12 schools in China and discuss the challenges faced by the different parties in the local school education sector for implementing and improving CTE.

The fourth chapter in this part—"Computational Thinking and the New Curriculum Standards of Information Technology for Senior High Schools in China" by Huang, Yang, Xiao, and Zhang—further shares a focused illustration of the new national curriculum standards designed and deployed for CTE in senior high schools in China. The authors introduce the evolution of new standards of the senior high school curriculum of information technology in China, review how CT relates to the key literacy emphasized in those new information technology curriculum standards, and discuss the deployment of teaching methods and the provision of learning environments favorable for students' CT development.

In "Computational Thinking Curricula in Australia and New Zealand," Bell, Vivian, and Falkner provide a holistic review and discussion of how Australia and New Zealand evolve and implement their national policies and curricula for CTE in K–12. The authors systematically depict and compare eight important aspects of the planning and implementation of CTE under the digital technologies subject in the two neighboring countries. This chapter serves as an informative guide for readers on the national plans and efforts for CT curricula in Oceania countries—with an inherent concern for equity, diversity, and inclusion in their curriculum efforts for CT development.

The final chapter in this part—"Computational Thinking Assessment: A Developmental Approach" by Román-González and Pérez-González—shares insights into the practices and instruments for the developmental approach to CT assessment, an important part of CT curricula in K–12 schools. The authors articulate and illustrate the rationales, targets, and approaches for the assessment of CT development at four schooling stages—kindergarten, elementary school, middle school, and high school. This chapter provides a theory-based guide for international readers on the design and implementation of longitudinal assessment of CT development among students.

PART II: SCHOOL IMPLEMENTATION AND CLASSROOM PRACTICE OF COMPUTATIONAL THINKING CURRICULA IN K–12

The second part of this book shares inspiring cases of the school implementation and classroom delivery of CT curricula in K–12. Seven chapters contributed by leaders in CTE in Asia and Europe present current conditions and research findings in relation to this focus area.

"Pathways of Computing Education: Formal and Informal Approaches" by Looi, Chan, Seow, Wu, and Wadhwa shares and illustrates the differentiated pathways of CTE implementation in formal versus informal learning systems in K–12 schools in Singapore. The chapter shares the experience in another important hub of e-learning in the Asia-Pacific region on the nationwide directions and efforts for CTE in K–12 through in-class and after-school initiatives. The authors first take a global angle to review the affordances of and synergies between formal and informal learning approaches to CTE; they then take a nation-specific focus to share the rationale and progress of CTE in Singapore through formal curricular and informal learning initiatives.

The second chapter in this part—"Software Education in South Korea for Cultivating Computational Thinking: Opportunities and Challenges" by So, Kim, and Ryoo—discusses and illustrates the current status, opportunities, and challenges of curricula implementation for CT development in K–12 schools in South Korea. This chapter presents readers with a reflective sharing from a technology hotspot in East Asia on the nationwide curricular initiatives for CT development under the policy on software education in K–12. The authors discuss the three unique features in the curricular directions and initiatives for software education in South Korea; they then reflect on the three major barriers and corresponding opportunities in relation to the capacity and perception of local K–12 teachers for software education.

In "Learning Computational Thinking in Co-Creation Projects with Modern Educational Technology: Experiences from Finland," Mäkitalo, Tedre, Laru, Valtonen, Iwata, and Koivisto discuss the cases of curriculum integration and teacher development for CTE in school education in Finland. This chapter shares inspiring experiences of CTE in a Nordic innovation hub whose education system is internationally well-recognized. The authors focus their sharing on two large-scale CTE initiatives—one about the school-based supports for fostering students' CT development through a game-based learning approach in the nonformal educational context, and the other about the national-level laboratory for preparing teachers' capabilities and beliefs for integrating CTE into everyday formal curricula.

"Integrating Design Thinking into K–12 Computational Thinking Classrooms in Taiwan: Practices of Collaborative Robotic Projects" by Shih gives a case of classroom practice in K–12 schools in Taiwan on students' synergic development of CT and design thinking. This chapter presents readers with an inspiring idea from another technology hotspot in East Asia for instructional innovation in CTE. The author introduces the theoretical background and learning model of "Situative CT" for instructional innovation in line with the existing standards and structures of Taiwanese school curricula and shares the ways and results of implementing instructional innovation in local K–12 classrooms for students to develop CT and design thinking in an integral manner.

The fifth chapter in this part—"Plethora of Skills: A Game-Based Platform for Introducing and Practicing Computational Problem Solving" by Armoni, Gal-Ezer, Harel, Marelly, and Szekely—shares a case on the experience and outcome of using a game-based learning platform in school education in Israel to develop students' skills of computational problem-solving. This chapter introduces readers to a leading tool for CTE implementation in an innovation hotspot in the Middle East. The authors articulate the philosophical principles and tool features in the design of the game-based learning platform and illustrate the use of the game-based learning platform under the novel pedagogical paradigm of "scenario-based programming" for CT development in classroom-level teaching and school-level competitions.

In "TA to AI: Tinkering Approach to Artificial Intelligence and Computational Thinking Education in Indian Schools," Raina, Jogeshwar, Yadav, and Iyer present a case of CT development through artificial intelligence (AI) education—a focus direction of CTE in K–12 in the coming decade. This chapter shares a stimulating achievement by a technology hotspot in South Asia on the development of national initiatives and materials for a tinkering curriculum in CTE in the big data era. The authors introduce the evolution and rationale of the AI education through tinkering approach in the Indian K–12 sector, and they illustrate the latest government–academy collaboration on developing guidelines, curriculum modules, national labs, and pedagogical resources for AI education at the nationwide level.

The final paper in this part—"Case Studies of Computational Thinking Education and Robotics Education in China" by Wang and Li—shares and discusses the experience in China on the curriculum planning and implementation of robotics education for CT development through physical computing education—another focus direction of CTE in K–12 in the coming decade. This chapter guides readers to take a look at the strategic planning and curricular initiatives for CT development through robotics education among K–12 students in China. The authors elaborate on the policy background, curriculum rationales, and syllabus structures in China for CT development through robotics education in primary and secondary schools, with a discussion about the cases of in-school and after-school robotics education activities for CT development in K–12.

THE INSIGHT INTO THE GLOBAL DIRECTIONS OF COMPUTATIONAL THINKING CURRICULA IN K–12

This section—taking the Darmstadt Model (Hubwieser 2013; Raman, Venkatasubramanian, Achuthan, and Nedungadi 2015) into consideration—grounds the chapters within a unifying framework for a critical highlight of visions, challenges, and initiatives in the development and implementation of CT curricula on a global basis. Table I.1 provides a critical highlight of the main issues in the development and implementation of CT curricula across the Asian, European, and Oceanian countries/regions covered by this volume.

Table I.1 The state of CT curricula implementations across the Asian, European, and Oceanian countries/regions covered by this volume

Countries/Regions (in alphabetical order)	Main issues in the development and implementation of CT curricula
Asia China *[CTE for primary and secondary school sectors (equivalent to grades 1 to 12)]*	• CT development is built through information technology education. • An official strategy with CTE-focused concerns was released in 2012. • CT is a noncompulsory curriculum component—allowing flexibility for the development and implementation of CTE-related initiatives at statewide, citywide, or school-based levels. • Inquiry-based learning for problem-solving is emphasized in the design of CTE-related initiatives. Unit-based teaching is advocated for CT cultivation through programming education and robotics education.
Hong Kong *[CTE for senior primary school sector (equivalent to grades 4 to 6)]*	• CT development is built through the coding education track under STEM education. • An official strategy with CTE-focused concerns has been released since 2015. • CT is a noncompulsory curriculum component—allowing school-based development and implementation of CTE-related initiatives. • Student-centered learning is emphasized for the design of CTE-related initiatives. Theme-based teaching is advocated for implementing CTE-related initiatives.
India *[CTE for upper primary to senior secondary school sectors (equivalent to grades 6 to 12)]*	• CT development is built through computer education and AI education. • An official strategy with CTE-focused concerns has been released since 2016. • CT is an elective component in the higher secondary curriculum and a noncompulsory component in the primary and middle school curriculum; yet it is an independent module in a national AI-education scheme across grades 6 to 12. • Active learning for problem-solving is emphasized for the design of CTE-related initiatives. An experiential teaching approach in authentic problem-based learning tasks is advocated for the implementation of CTE-related initiatives.
Israel *[CTE for elementary, middle, and high school sectors (equivalent to grades 1 to 12)]*	• CT development is built through the computer science program. • An official strategy with CTE-focused concerns has been released since 2012. • CT is a compulsory component in the high school curriculum and a noncompulsory component in the elementary and middle school curriculum. • The ability of computational problem-solving is the emphasis on CTE-related initiatives. A scenario-based programming approach is advocated for the cultivation of computational problem-solving.

(continued)

Table I.1   (continued)

Countries/Regions (in alphabetical order)	Main issues in the development and implementation of CT curricula
South Korea *[CTE for upper elementary, middle, and high school sectors (equivalent to grades 5 to 12)]*	• CT development is built through the *software education* curriculum. • An official strategy with CTE-focused concerns has been released since 2014. • CT is a compulsory curriculum component—as in the *software education* subject for grades 5 to 12. • Software education is the main approach to CTE, which integrates with science, technology, engineering, arts, and mathematics (STEAM) education and maker education for an emphasis on students' development of CT together with creativity and communication skills.
Singapore *[CTE for primary and secondary school sectors (equivalent to grades 1 to 12)]*	• CT development is built through computing/computer science education. • An official strategy with CTE-focused concerns has been released since 2014. • CT is a noncompulsory curriculum component—offering CTE-related initiatives to students at various ages through various touchpoints on campus and after school. • The approaches of inquiry-based and project-based learning for problem-solving are advocated for the design and implementation of CTE-related initiatives in a school-based manner.
Taiwan *[CTE for primary and secondary school sectors (equivalent to grades 1 to 12)]*	• CT development is built through the *information technology* track and *life and technology* track under the technology learning area of the national curriculum. • An official strategy with CTE-focused concerns has been released since 2014. • CT is a noncompulsory curriculum component—school-based approaches to integrating CTE-related initiatives into cross-field learning with real-life scenarios. • Situated learning through problem-solving tasks is emphasized for the design of CTE-related initiatives. A theme-based teaching approach is advocated for the implementation of CTE-related initiatives.
Europe Finland *[CTE for basic education sector (equivalent to grades 1 to 9)]*	• CT development is built through the *computing* cross-curricular topic, which is integrated in all subjects in the national core curriculum for basic education. • An official strategy with CTE-focused concerns has been released since 2014. • CT is a compulsory curriculum component—as in the *computing* cross-curricular topic for grades 1 to 9. • The design and implementation of CTE-related initiatives emphasizes the human-driven, design-driven, and discipline-driven perspectives for a holistic understanding of CT.

Table I.1 (continued)

Countries/Regions (in alphabetical order)	Main issues in the development and implementation of CT curricula
United Kingdom *[CTE for primary and secondary school sectors (equivalent to grades 1 to 12)]*	• CT development is built through computing/computer science education. • An official strategy with CTE-focused concerns has been released since 2012. • CT is a compulsory curriculum component in England and Scotland but a noncompulsory curriculum component in Wales and Northern Ireland. • Problem-solving elements is emphasized in the design and implementation of CTE-related initiatives.
Oceania Australia *[CTE for primary and secondary school sectors (equivalent to grades K to 10)]*	• CT development is built through the *digital technologies* curriculum. • An official strategy with CTE-focused concerns has been released since 2011. • CT is a compulsory curriculum component—as in the *digital technologies* subject for grades K to 10. • Both plugged and unplugged approaches are emphasized in the design and implementation of CTE-related initiatives.
New Zealand *[CTE for primary and secondary school sectors (equivalent to grades K to 12)]*	• CT development is built through the *digital technologies* curriculum. • An official strategy with CTE-focused concerns has been released since 2011. • CT is a compulsory curriculum component—as in the *digital technologies* subject for grades K to 12. • Both plugged and unplugged approaches are emphasized in the design and implementation of CTE-related initiatives.

School education in countries around the world calls for the cultivation of students' CT, with a vision that CT is an essential skill for everyone in the digital society. Between 2011 and 2016, CTE started to become a globally emergent focus in curriculum planning, development, and implementation in K–12 schools. The countries/regions covered by this volume document official strategies for CTE mostly across grades 1 to 12. For the Oceanian countries of Australia and New Zealand, CT curricula start as early at grade K. In Hong Kong, South Korea, and India, the introduction of CTE elements into school curricula starts in senior primary schooling, in grades 4, 5, and 6, respectively.

Almost all countries/regions covered by this volume take the existing curricula in the disciplines related to computing, computer science, information technology, and digital technologies as a foundation when designing K–12 curricular initiatives for CT development. This direction in curriculum design leads to the popularity of the pedagogical emphasis on the problem-based learning approach for problem-solving in CTE, as the goal of these disciplines commonly targets the use of digital technologies to solve problems. Theme-based inquiry tasks are commonly arranged in CTE-related initiatives in this regard. Israel makes a noteworthy innovation of using the scenario-based programming approach in CTE-related initiatives for the cultivation of computational problem-solving.

Nearly half of the countries/regions covered by this volume do not require K–12 schools to compulsorily deliver subject curricula for CTE. For the remaining countries/regions, the governments set CT to be a compulsory component in the school education curriculum. In the United Kingdom, CT is a compulsory curriculum component in two of its constituting nations: England and Scotland. In Israel, CT is a compulsory component in the high school curriculum, not in the elementary and middle school curriculum. In Finland, South Korea, Australia, and New Zealand, schools are required to deliver CT-related curricula components in all schooling grades covered by the national CTE policy: the *computing* cross-curricular topic with a human-driven, design-driven, and discipline-driven concern in the Finnish core curriculum for basic education; the *software education* subject in the South Korean national curriculum for developing CT, creativity, and communication in upper elementary, middle, and high school sectors; and the *digital technologies* subject with plugged and unplugged activities in primary and secondary curricula in Australia and New Zealand.

This volume provides insights into the promising directions of computational thinking curricula in K–12. The curricular implementation of CT education in school education should be started as early as at the kindergarten level, for providing students with a coherent development of CT throughout their four schooling stages—kindergarten, elementary school, middle school, and high school. Students at the kindergarten stage are able to enjoy CT education through symbolic plays that involve the prerequisites of CT (decomposition and pattern recognition), based on the discussion by Román-González and Pérez-González.

The experience in the United Kingdom (as shared by Scott), New Zealand, and Australia (as shared by Bell, Vivian, and Falkner) as well as China (as shared by Yang and Ren) demonstrates the feasibility for policymakers and school leaders to plan and implement CT curricular initiatives covering grades K through 12. In China, there is a revamp of national curriculum standards for implementing CTE (as shared by Huang et al.).

From the sharing among all Asian, European, and Oceanian countries/regions covered by this volume, the stages of elementary schooling and middle schooling are considered to be a natural timepoint for promoting students' development of CT in school education; and the approach of problem-solving in the theme-based inquiry process when learning computer programming is popularly chosen for CT curricular activities to develop students' CT. These CT curricular focuses concur with the global meta-review findings in Grover and Pea (2013) and Hsu et al. (2019).

Apart from the stream of programming education, this volume presents various growing concerns for CT curriculum implementations: CT development through robotics education (experience in China as shared by Wang and Li); CT development through AI education (experience in India as shared by Raina et al.); CT development through STEAM education (experience in South Korea as shared by So, Kim, and Ryoo); CT development with an emphasis on human-driven, design-driven, and discipline-driven perspectives (experience in Finland as shared by Mäkitalo et al.).

Most countries/regions set CT as a compulsory component in the school curriculum, while three Asian countries/regions covered by this volume—namely Hong Kong (as shared by Kong and Kwok), Singapore (as shared by Looi et al.), and Taiwan (as shared by Shih)—opt to arrange CT to be a noncompulsory curriculum component. Schools in these three places are allowed school-based flexibility to develop and implement CTE-related initiatives on campus and after school.

On top of sharing global experience in CT curricula implementations with the conventional focuses, this volume also provides an alternative perspective for curriculum developers and school practitioners to diversify the pedagogical designs for CTE integration in K–12 school curricula. The Israeli experience, as shared by Armoni et al., reframes CT to be "computational problem-solving" and innovates the pedagogical paradigm of "scenario-based programming" for CT development. Curriculum developers and school practitioners are encouraged to think outside the box to go beyond the focus on imperative sequential programming when they design and implement CT curricular initiatives in school education.

In summary, policymakers, school leaders, and school teachers in K–12 sectors around the world are creating promising change at the levels of education system, school implementation, and classroom practice for the planning, development, and implementation of CT curricula. The goal of this volume is to inform and advance debate and discussion about CT curricula in K–12. This volume organizes thirteen chapters into two parts to broadly address four areas of concern in CT curricula implementations— educational policy for CTE in K–12, curriculum development for CTE in K–12, school implementation of CT curricula in K–12, and classroom practice of CT curricula in K–12. The global and representative overview provided in this volume can provide a quick reference for policymakers, curriculum developers, school practitioners, and educational researchers to develop an international perspective of and a rounded reflection on the current states and future possibilities of CT curricula in K–12 schools.

REFERENCES

Borrás, Susana, and Charles Edquist. 2015. "Education, Training and Skills in Innovation Policy." *Science and Public Policy* 42, no. 2: 215–227.

Grover, Shuchi, and Roy Pea. 2013. "Computational Thinking in K-12: A Review of the State of the Field." *Educational Researcher* 42, no. 1: 38–43.

Hsu, Yu-Chang, Natalie Roote Irie, and Yu-Hui Ching. 2019. "Computational Thinking Educational Policy Initiatives (CTEPI) across the Globe." *TechTrends* 63, no. 3: 260–270.

Hubwieser, Peter. 2013. "The Darmstadt Model: A First Step Towards a Research Framework for Computer Science Education in Schools." In *Informatics in Schools. Sustainable Informatics Education for Pupils of all Ages. ISSEP 2013. Lecture Notes in Computer Science, Vol. 7780*, edited by Ira Diethelm and Roland T. Mittermeir, 1–14. Berlin, Heidelberg: Springer.

Mao, Jin, Dirk Ifenthaler, Toru Fujimoto, Andrea Garavaglia, and Pier Giuseppe Rossi. 2019. "National Policies and Educational Technology: A Synopsis of Trends and Perspectives from Five Countries." *TechTrends* 63, no. 3: 284–293.

Raman, Raghu, Smrithi Venkatasubramanian, Krishnashree Achuthan, and Prema Nedungadi. 2015. "Computer Science (CS) Education in Indian Schools: Situation Analysis Using Darmstadt Model." *ACM Transactions on Computing Education* 15, no. 2: 1–36.

Williamson, Ben, Annika Bergviken Rensfeldt, Catarina Player-Koro, and Neil Selwyn. 2019. "Education Recoded: Policy Mobilities in the International 'Learning to Code' Agenda." *Journal of Education Policy* 34, no. 5: 705–725.

I

EDUCATIONAL POLICY AND CURRICULUM DEVELOPMENT FOR COMPUTATIONAL THINKING EDUCATION IN K–12

1

COMPUTATIONAL THINKING EDUCATION IN THE UK: A TEACHER'S PERSPECTIVE

Jeremy Scott

THE UNITED KINGDOM

The United Kingdom of Great Britain and Northern Ireland (UK) is a unitary parliamentary democracy and constitutional monarchy comprising four nations—England, Scotland, Wales, and Northern Ireland—and four education systems. England is the largest country both by land area and population (population 53 million) followed by Scotland (population 5.3 million), Wales (population 3.1 million), and Northern Ireland (population 1.8 million) (UK Office for National Statistics 2011). Scotland, Wales, and Northern Ireland have devolved governments,[1] responsible for a range of policies, with the UK government reserving powers over areas such as macroeconomic policy and defense.

Education is a devolved matter and while each nation has its own school education system, there are commonalities across all four. The majority of children attend state-funded comprehensive schools from age five to eighteen, while Northern Ireland retains state-funded grammar schools for approximately 45 percent of students who pass a selection examination at age eleven (Meredith, 2018). Around 6.5 percent of the total number of school children in the UK attend independent (normally fee-paying) schools (Independent Schools Council 2018).

Although the age ranges vary slightly between the four nations, the primary phase generally lasts from around five to twelve years of age and the secondary phase from about twelve to eighteen years of age.

Students in England, Wales, and Northern Ireland sit for the same public examinations (General Certificate of Secondary Education (GCSE) at around age sixteen and A-level at around age eighteen), while Scotland has a distinctive system of education and public examinations (taken from about ages sixteen to eighteen) that is separate from the rest of the UK.

TERMINOLOGY USED IN THIS CHAPTER

COMPUTATIONAL THINKING

Computational thinking is an approach to solving problems that involves viewing the world through thinking practices derived from computer science (CS). These practices can be grouped into four main areas:

- seeing a problem and its solution at multiple levels of detail (abstraction)
- thinking about problems as a series of steps (algorithms)
- understanding that solving a large problem will involve breaking it down into a set of smaller problems (decomposition)
- identifying similarities and repetition within and across problems (pattern recognition)—and realizing that a solution to a problem may be used or adapted to solve related problems

Underpinning this are some key understandings about computers:

- Computers are deterministic—they do what you tell them to do.
- Computers are precise—they do exactly what you tell them to do.
- Computers can therefore be understood—they are just machines with logical working.

COMPUTER SCIENCE

Computer science is an academic discipline and school subject that deals with problem-solving through the use of algorithms and data structures to design and create computer programs. While computational thinking can be applied in many school subjects and areas of life, computer science[2] is particularly well placed to deliver it.

COMPUTING
In an educational context, this is a broad term used to describe all computer-based education in schools including digital literacy, information and communication technology (ICT) (see below), and computer science.

DIGITAL LITERACY
Digital literacy refers to the ability to use a computer effectively as a tool for learning as well as creating and presenting work across the curriculum. Typically, this involves using the Internet for research and applications packages to present work.

INFORMATION AND COMMUNICATION TECHNOLOGY
Information and communication technology is a school subject that involves the teaching of core office applications[3]—word processing, spreadsheet, email, and presentation software—largely to secondary school students.

WHAT IS THE PURPOSE OF EDUCATION?
We are in the midst of a technological and societal revolution that is effecting faster and greater change to society than at any period in history. No aspect of our lives will be left untouched—and at its core is information and computation.

Most would agree that the purpose of education is to produce literate, numerate, and knowledgeable individuals who are ready to take their place in a country's workforce. Today's children will need the knowledge and skills to undertake jobs that will require, or at least benefit from, an understanding of computers and the software that drives them. At a societal level, the education system will provide computing experts to supply an increasingly knowledge-based economy; at an individual level, it will enable young people to gain skilled, high-quality, and well-paid employment.

Education, however, must also be about a deeper, individual benefit, enabling children to develop the knowledge, skills, and abilities they will need to make sense of the world around them. This was as true 250 years ago for Rousseau's *Emile*, educated to navigate his way through the perils of human society and take his place in it (Rousseau 1762), as it is for today's children navigating their way in a society with information and computation at its core.

Children learn about mathematics, science, and history not because they will become scientists, mathematicians, or historians: only very few actually will; rather, they study these subjects—and others—to make sense of the world around them and acquire agency. Someone who therefore does not understand computation and information will not be equipped to cope with the successive waves of technological change that they will live through. In short, they will be unable to participate fully in society. Instead, they will see technology as something mysterious of which they have little understanding and will not be equipped to make informed judgments on technological developments that will affect their lives.

This is why computational thinking must be a foundational discipline, at the core of every modern education system, just like mathematics, science, and humanities. It is this belief that has driven the changes seen in UK computing education in the twenty-first century.

A BRIEF HISTORY OF COMPUTING IN THE UK

The UK has a long computing history, going back to Charles Babbage in the mid-nineteenth century and Ada Lovelace, often regarded as the world's first computer programmer. The UK can also claim to be the country that invented the computer, in the shape of Colossus—the world's first programmable electronic digital computer—that helped Allied code-breakers to break Nazi ciphers during World War II.

Fast-forward to the early 1980s, however, and the UK had lost its lead, with the US's IBM dominating the first computer revolution, then companies such as Apple and Microsoft leading the second, personal computer, revolution. Enter the UK's public service broadcaster, the BBC, with its mission statement to "inform, educate and entertain." In a tie-up with emerging computer manufacturer Acorn,[4] the BBC sought to introduce the public to computers and show what they were capable of in what would become the BBC Computer Literacy Project. An accompanying national television series *The Computer Programme* began in 1982, showcasing different uses of computers and, importantly, featured projects that introduced the public to programming.

The resulting BBC Microcomputer was, thanks to a UK government drive, adopted by around 80 percent of UK schools (Vasko and Dicheva 1986, 7) and also found success in the home market. With little off-the-shelf application software available, a built-in BASIC programming language interpreter encouraged a generation of school children to learn to program to make it do what they wanted and is credited with helping to kickstart computing education in UK schools (Arthur 2012).

The arrival in schools in the late 1980s of more powerful computers with a range of productivity software available for them largely ended the era of self-made classroom programmers; the emphasis with the new machines was use of office applications rather than programming.[5] Computing education had become about the technology itself rather than the foundational intellectual discipline of computational thinking—creating a generation of tool *users* rather than tool *creators*.

WHAT WAS HAPPENING IN EACH NATION OF THE UK IN RESPECT OF CT AND CS

By the start of the twenty-first century, school children across the UK received a varying amount of computing education, which could broadly be divided into three areas: digital literacy, ICT, and computer science (The Royal Society 2012, 5).

At this time there was no rigorous primary school curriculum in computing; the level of exposure that children received in the primary classroom was often dependent on whether their teacher was a computer enthusiast. Where computers were used, the focus tended to be on digital literacy and enhancement of learning across the curriculum.

ENGLAND, WALES, AND NORTHERN IRELAND

In England, Wales, and Northern Ireland, ICT was the mainstream offering in secondary schools, rooted in office applications. Teachers delivering the courses frequently commented that students found this boring and contrasted with the increasingly rich multimedia applications they used at home. While born out of a well-intentioned initiative to create a generation of school leavers equipped with the skills it was perceived they would require in a modern workplace, the subject degenerated into the teaching of outdated skills by teachers, the majority of whom did not hold a relevant qualification (The Royal Society 2012, 72).

CS remained largely a *niche* subject. Many schools did not offer it and among those that did, it was offered from age fourteen and up. It saw relatively low uptake and gradually decreasing numbers (The Royal Society 2012). It was a far cry from being a foundational subject or cornerstone of a modern curriculum.

SCOTLAND

Things were different in Scotland, where there had been a well-established tradition of CS education in secondary schools since the 1980s. Computing studies[6] was a popular choice for students selecting subjects for public examinations from the age of fourteen, and, despite having around one-tenth of the population of England, Scotland could boast of having broadly similar numbers of students taking CS in the main public examinations (Scottish Qualifications Authority n.d.).

Scottish secondary school teachers are also required to have a university background and teaching qualification in their specialism, so students studying CS benefited from being taught by qualified subject specialists. CS had been seen as an ideal vehicle to deliver digital literacy, however; consequently, courses had become broader, and the number of students choosing the subject had started to decline.

BEGINNING OF THE GROUNDSWELL

By the early twenty-first century, the situation was therefore ripe for a reboot of the computing curriculum throughout the UK.

Computational thinking paper by Jeanette Wing   While the phrase *computational thinking* was first used by Seymour Papert in 1980 (Papert 1980, 182), Wing's 2006 paper popularized it and focused minds around what school computing could and should be about. CT also portrayed CS as an intellectual discipline concerned with the type of higher-order thinking identified by Piaget and others (Donaldson 1978) in contrast to what some viewed as the lower-order skills currently being taught. It could be seen that the ability to think computationally would be advantageous to all twenty-first-century citizens and that any country serious about the future of its learners and economy must get behind CT education in its schools.

Eric Schmidt's comments at the MacTaggart Lecture in 2011 Other publications that followed bemoaned the lack of CT in the UK's schools. It was Eric Schmidt (as chairman of Google), however, who arguably made a bigger impact when he criticized the state of computing education in the UK during his MacTaggart lecture at the 2011 Edinburgh Festival:

We need to reignite children's passion for science, engineering and maths. In the 1980s the BBC not only broadcast programming for kids about coding, but (in partnership with Acorn) shipped over a million BBC Micro computers into schools and homes. That was a fabulous initiative, but it's long gone.

I was flabbergasted to learn that today computer science isn't even taught as standard in UK schools. Your IT curriculum focuses on teaching how to use software but gives no insight into how it's made. That is just throwing away your great computing heritage.

Schmidt not only laid bare a problem with UK education but dented the pride of a nation that was the land of Babbage and Turing and had created the first electronic computer. As a senior industry figure, his intervention also grabbed media headlines—and the attention of policymakers.

THE WATERSHED MOMENT

Shutdown or restart? At the same time, The Royal Society of London (the UK's national academy of sciences) was investigating computing education in the UK to suggest the best way forward. The publication in 2012 of *Shut Down or Restart? The Way Forward for Computing in UK Schools* was a watershed moment that challenged those in education and government to improve the teaching of computing in schools. Its very title—*Shutdown or Restart?*—suggested that pulling the plug on CT education was a serious prospect.

Published thirty years after the launch of the BBC Microcomputer, it painted a picture of missed opportunities and wrong turns in computing education that had inadvertently led us to the point of near extinction of CS as a school subject, in the quest to create a digitally literate population.

It made several recommendations, including calling for a review of curricula and qualifications, recruitment and training of specialist teachers, provision of continuing professional development (CPD), and better links between school and university courses. Most of all, its principal thrust was to move from teaching tool *use* to tool *creation* through the foundational discipline of CT.

Computing had changed   These developments were happening against a backdrop of enormous change in computer technology. With the continuing expansion of the World Wide Web and the development of smartphones and tablet computers, computing was becoming increasingly mobile, personal, and more pervasive than ever. Everyone used computers on a daily basis, and industry was desperate for more qualified computing professionals, especially programmers.

More accessible tools: Blocks-based programming   Furthermore, the tools that educators had at their disposal were changing. The introduction of block-based programming languages, exemplified by MIT's Scratch and App Inventor, made the teaching of programming easier and more accessible than ever before. Anyone who has tried to teach children how to program will know that the first—and for many learners the biggest—hurdle can be working with syntactically demanding, text-based programming languages. Those learners reluctant to persevere, or who have not had a positive experience of education to date, may fall at this first hurdle.

These new environments, on the other hand, allow learners to focus on problem-solving in a way that removes those barriers and engages them. They provide an accessible introduction to programming that allows quick results in a rich, multimedia environment that encourages creativity and exploration.

Political recognition of the importance of CS   So, the stars were aligned and the gauntlet thrown down. It was up to those in education and government to rise to the challenge and make CT education, particularly through the study of CS, an integral part of a twenty-first-century curriculum.

INITIATIVES

ENGLAND
A new national curriculum   In 2011, the UK government announced a review of England's national curriculum. As part of that review, the government asked two learned societies—the British Computer Society (BCS) and

the Royal Academy of Engineering—to draft a new computing curriculum. In 2014, a new national curriculum in computer science was introduced that aims to ensure that all children from ages five through fourteen:

- can understand and apply the fundamental principles and concepts of computer science, including abstraction, logic, algorithms and data representation
- can analyze problems in computational terms and have repeated practical experience of writing computer programs in order to solve such problems
- can evaluate and apply information technology, including new or unfamiliar technologies, analytically to solve problems
- are responsible, competent, confident, and creative users of information and communication technology

In secondary schools, the subject of ICT was to be replaced by computer science. ICT teachers—many of whom had been the school computer enthusiast who had taken on the mantle of teaching the subject—would now be required to deliver a curriculum in which the large majority held neither a relevant degree nor teaching qualifications (The Royal Society 2012, 72).

Computing education would be compulsory and would no longer focus on the technology itself, but on the foundational discipline of CT from the early years to secondary school. Such a bold move, however, presented a huge challenge in terms of support for upskilling many nonspecialist teachers in subject knowledge as well as pedagogy.

Computing at School   Computing at School (CAS)[7] was formed in 2008 as a volunteer-driven grassroots organization that sought to provide support for teachers trying to deliver the new CS curriculum in schools across England and Wales. Adopting a mantra of "There is no 'them,' only us," CAS's declared aim is for "computing with computer science at its heart, to become firmly established in all primary and secondary schools, alongside mathematics and the natural sciences and to support all teachers committed to providing high quality computing education for their students" (Computing at School 2019).

With support from the government and industry, CAS set about building a "network of teaching excellence," whereby "master teachers" would receive training to form and lead local hubs with support from the BCS and computing professionals, including university CS departments. Major IT companies, as well as smaller organizations, provided professional "ambassadors" on a *pro bono* basis.

Computing at School has been a highly successful organization that could be described as an enormous self-help group. Self-sustaining, it remains faithful to its grassroots origins and is now recognized as the official subject association for CS in UK schools.

At the time of writing, CAS has a worldwide membership of over thirty thousand (including the author). Its five hundred master teachers have set up 254 local teacher hubs, and it has delivered more than 4,500 hours of professional learning in the past year. CAS also hosts more than four thousand online resources. Building such a community of good practice will doubtless act as a template for success in other education systems.

Barefoot—A primary curriculum In 2014, the BCS set up the Barefoot program with England's Department for Education. Initially funded as a one-year program to help primary teachers in England prepare for the changes in the computing curriculum, Barefoot found early success and funding was taken over by national telecoms company BT in 2015 (which continues to fund it and runs it in partnership with CAS).

Barefoot provides CS teaching and learning resources for primary teachers with no prior expertise; these are designed to be cross-curricular, and most are unplugged[8] to be as accessible as possible. Like CAS, Barefoot adopts a grassroots model, with all content developed by teachers. The project also provides CPD workshops for schools that are delivered by volunteer IT professionals. All of this is free of charge to primary schools across the UK.

Barefoot now has widespread adoption across the UK and a reach beyond its borders. To date, more than seventy thousand primary teachers across the UK have benefited from the project, with 95 percent saying that the professional development workshops improved their confidence and ability in the classroom (Barefoot Computing 2020).

National Centre for Computing Education In 2019, the National Centre for Computing Education (NCCE) was set up to work with schools across England to improve the teaching of computing and drive-up participation in post-14 CS courses leading to public examinations.

A consortium made up of STEM Learning,[9] BCS, and the Raspberry Pi Foundation[10] is delivering the work of the NCCE, backed by up to £84 million of government funding.

Chaired by computer scientist Simon Peyton Jones, the NCCE operates virtually through a network of up to forty school-led computing hubs[11] to provide training and resources to primary and secondary schools and an intensive training program for secondary teachers without a post-high-school qualification in computer science. The NCCE works with the University of Cambridge, with a further £1 million investment from Google. Peyton Jones describes it as "the biggest single per capita investment in teacher professional development for computing education anywhere on the planet." Indeed, by late 2020, more than 1,300 teachers had completed the NCCE's Computer Science Accelerator program, ready to teach GCSE Computer Science (National Centre for Computing Education 2020).

SCOTLAND

Curriculum for Excellence   "North of the border" in Scotland, things were moving apace. The Scottish government was in the process of a root-and-branch revision of the entire school curriculum and following a lengthy consultation and development period, the new *Curriculum for Excellence* was implemented in 2010.

CS was introduced for all learners from age five to fifteen by adopting three strands:

- understanding the world through computational thinking
- understanding and analyzing computing technology
- designing, building, and testing computing solutions

Like initiatives elsewhere in the UK, Scotland's new curriculum established CT as a foundational discipline from the primary phase (ages c. five to twelve) but also provided significant challenges in terms of CPD for teachers without experience of it.

National examinations   Following the introduction of the new curriculum, there was a wholesale revision of courses for public examinations. A new course emerged that amalgamated two former courses—Computing and Information Systems—into a single course called Computing Science that largely eschewed application package use in favor of a more formal CS approach through software, database, and web development. The new course runs at various levels of progression from ages fifteen to eighteen.

The Royal Society of Edinburgh Exemplification Project Such a radical change required support. With early secondary school (ages twelve to fifteen) in mind, the Royal Society of Edinburgh (RSE), Scotland's national academy of arts and sciences, launched a project to exemplify the teaching of CS both in Scotland and in the rest of the UK. I was recruited to create supporting materials, creating three packs in the first phase of the project (The Royal Society of Edinburgh 2016):

- *Starting from Scratch*: an introduction to CS using MIT's Scratch.
- *Itching for More*: an intermediate CS course using UC Berkeley's BYOB.[12]
- *I ♥ My Smartphone*: consolidating previous work through the medium of mobile app development using MIT App Inventor.

The packs comprise screencasts (tutorial videos) to illustrate program construction for the YouTube generation. A "core + extension" approach allows all learners to achieve success while stretching the more able or experienced, often luring them into making errors and challenging them to debug faulty code. In each lesson, embedded questions encourage discussion and collaboration between students as they seek to tease out misconceptions and reinforce deeper understanding by debugging code samples or examining concepts such as sequencing, scope, and hierarchy.

Mindful that in these earlier years of secondary school CS may be delivered by nonspecialists, the packs also include teacher materials containing pedagogical guidance. Interdisciplinary learning opportunities—a cornerstone of Scotland's new curriculum—are also signposted, encouraging teachers to step out of their comfort zones, while mapping to the curriculum's aims and objectives.

Computing at School Scotland Around the same time as the RSE project, Scotland's secondary school CS teachers became more organized, forming Computing at School (CAS) Scotland, linked to its counterpart in England. Working solely on a volunteer basis, CAS Scotland sought to support and provide a voice for secondary school CS teachers. It is true that CAS Scotland is less active than its counterpart in England, possibly because CS was already well established in Scotland with its own online support community CompEdNet[13] and because the pace and scope of radical curriculum reform made it difficult for its membership to commit as much time as many would have liked. Nonetheless, CAS Scotland has provided support for CS teachers, ensuring that their voices are heard during multiagency discussions, and has also organized well-received national conferences for CS teachers.

Following recommendations in the independent *Scottish Technology Ecosystem: Review*,[14] a new body called Scottish Teachers Advancing Computing Science (STACS)[15] has recently been set up to provide further ongoing support to Scottish CS teachers, with funding from the Scottish government (Scottish Government 2022).

PLAN C Despite Scotland's CS teachers being qualified in their subject, the new curriculum created a need for CPD for those delivering the courses. As CS had matured as a subject, much of the recent research into pedagogy and how children can learn to think computationally had not reached the classroom.

Consequently, in 2013, PLAN C (Professional Learning and Networking for Computing),[16] a two-year CPD program for secondary CS teachers, was developed in collaboration with the BCS and CAS Scotland, assisted by funding from the Scottish government. The existing subject expertise of Scottish CS teachers allowed the program's creators to focus on pedagogy. Echoing the model adopted by CAS in England, lead teachers were identified who received university training from the project leaders in the latest research as well as use of materials developed by the project staff. These lead teachers returned to their local hubs, delivering training to their colleagues in structured CPD meetings. Time was also set aside at hub meetings for staff to discuss and share classroom experiences—the project's creators considered high-quality professional dialogue to be a vital ingredient of the program—and the collegiate networking that resulted from this was seen as key to the project's sustainability.

By the end of the project, approximately half of Scotland's secondary school CS teachers had received PLAN C training, and the new pedagogy was having a beneficial effect on classroom practice (Cutts, Quintin, Robertson, Donaldson, and O'Donnell 2017, 47–48).

WALES

In 2008, the National Curriculum for ICT was launched in Wales. The curriculum applied to ages seven to sixteen and required students to develop skills in digital literacy, ICT, and data handling, but there was no requirement to deliver computational thinking.

A new Digital Competence Framework was therefore put in place in 2016, setting out the digital skills to be attained by learners aged between three and sixteen across four strands: citizenship; interacting and collaborating; producing; data and computational thinking.

In 2022, a new national curriculum was launched in Wales that will be introduced to nursery/kindergarten through to year seven (ages three to eleven) and rolled out to years eight to eleven (ages twelve to fifteen) between 2023 and 2026.

The new curriculum contains six "areas of learning and experience" (AoLE), each containing a number of "statements of what matters [for learning]." The science and technology AoLE includes the statement that "computation is the foundation for our digital world" and part of this—programming fundamentals—sees the importance of understanding the fundamentals of computation and is likely to be supported by a range of unplugged activities. Another part—implementation—will see students use programming languages to solve problems after having understood the fundamentals of computation.[17]

Like England and Northern Ireland, public examinations in CS are offered in Wales but not in every school.

NORTHERN IRELAND

Prior to 2016, the computing curriculum in Northern Ireland focused on the use of computers, much as in England, with relatively little CS (Perry 2015). In late 2016, however, the education minister for Northern Ireland announced the adoption and integration of the Barefoot program into Northern Ireland's curriculum (*Belfast Telegraph* 2016).

This had a major impact, according to Dr. Irene Bell, head of STEM at Stranmillis University College, one of Northern Ireland's initial teacher education institutions[18]. For the first time, CT was explicitly mentioned in the Northern Irish curriculum, although unlike England and Scotland, where CT and CS is a compulsory part of the curriculum, CT comes under the category of a "desirable" strand in the broader ICT curriculum.

Bell therefore undertook a project to map Barefoot resources to Northern Ireland's primary curriculum using terminology familiar to teachers there. Following this, a program of professional development was set up to train the nation's primary teachers in delivery of the Barefoot program. By the end of 2018, almost half of Northern Ireland's primary schools had participated in the training program.

Bell is realistic about the challenges that lie ahead: half of Northern Ireland's teachers still have to be trained, and there remain some hard-pressed primary teachers who struggle to see the importance of delivering computational thinking in an already crowded curriculum. She believes, however, that a focus on teacher training is vital and that with the support of high-quality materials from Barefoot, Northern Ireland is now getting teachers into primary schools who can deliver CT education.

In secondary schools, an ICT curriculum focuses on aspects of digital literacy up to age fourteen. In 2018, the Council for the Curriculum, Examinations & Assessment (CCEA) issued guidance for post-primary schools on delivering CT education, adapted from documentation produced by CAS in England (Berry 2015), although this remains guidance only and is not mandated (Council for the Curriculum 2018). The same public examinations in CS for students aged approximately sixteen and eighteen as those offered in England and Wales are available, but again they are not offered in every school.

UK-WIDE INITIATIVES

In addition to ongoing work by each of the four nations, there have also been several UK-wide initiatives, many supported by industry.

Micro:bit One example is the BBC Micro:bit, a small microcontroller computer developed by the BBC (in partnership with twenty-nine organizations across industry and education) with sensors such as an accelerometer, compass, and Bluetooth as well as a grid of programmable LEDs.

Invoking its public service role, and harking back to the days of the BBC Microcomputer, the BBC sought to "inspire a new generation to get creative with coding, programming and digital technology" with its development.

In 2016, a free Micro:bit was given to every eleven- to twelve-year-old in the UK, with extensive support materials for students and teachers to support CT education and create a standard platform for children and teachers both in the classroom and at home. Research commissioned by the BBC in 2017 on the impact of the initiative suggested that the project was widely regarded by teachers and students as successful (Discovery Research 2017).

Machine Learning for Kids The wave of machine learning services that has arrived in our lives prompted IBM UK developer Dale Lane to create *Machine Learning for Kids* in 2017. A free web-based service developed by Lane in his own time, it enables students to create machine learning models—of text, images, numbers, or sounds—that can be imported and used in Scratch, App Inventor, or Python. The site also contains worksheets and projects for teacher and student use and has achieved significant uptake: according to Lane, the system is used to create thousands of projects per month in the UK and hundreds of thousands worldwide.[19]

Not only does this allow students to gain insight into the field and make sense of how facial recognition or recommendations work on Spotify and TikTok but the introduction of machine learning into the classroom also raises questions about the kind of computational thinking that is taught in schools. Arising from this is the recent concept of "CT 2.0"—a more data-driven approach to problem-solving, in contrast to the current rule-driven "CT 1.0," suggesting how CT education must evolve as new paradigms in computing emerge (Tedre, Denning, and Toivonen 2021).

Charities Educational charities have also played an important role in the provision of CT education in schools, often in partnership with other organizations and industry. One example is the Raspberry Pi Foundation, which has been closely involved with a number of high-impact initiatives including:

- the Bebras Computing Challenge (in collaboration with the University of Oxford)—the UK arm of the popular computational thinking competition.[20]
- Code Club—a global network of free coding clubs for nine- to thirteen-year-olds run by volunteers and educators.[21]
- *Hello World* (in collaboration with Computing at School)—a free magazine published three times per year for UK-based computing educators to share experiences and learn from each other.[22]

CHALLENGES THAT LIE AHEAD

In 2017, the Royal Society published *After the Reboot*, an evaluation of the activity that had followed the publication in 2012 of *Shutdown or Restart?*. *After the Reboot* notes first and foremost that there is much to celebrate: many more children are receiving much deeper CT education than ever before (The Royal Society 2017, 5).

This provision, however, is patchy and fragile: only 11 percent of students take the GCSE, the main public examination at age sixteen in CS in England, Wales, and Northern Ireland (The Royal Society 2017, 7), and 54 percent of schools do not even offer the subject at public examinations (The Royal Society 2017, 45). Only one in five CS students is female (The Royal Society 2017, 7).

Continuing professional development is still inadequate with 25 percent of teachers lacking the opportunity to do any professional learning in CS and more than 40 percent who had done less than nine hours (The Royal Society 2017, 73).

Teacher education   Much has been done to establish CT education in schools, and it is now firmly part of the primary curriculum. This creates challenges, however—both for primary teachers stretched between curricular initiatives and school leaders trying to meet the demands of curricular requirements and school inspectors. Few teachers have studied CS at school, so one of the challenges is building an understanding in teachers of what CT actually is.

Furthermore, while CPD for existing teachers is vital—and much has been done to date—there is also a need for significant CT to be embedded in initial teacher education.

Computing is also a fast-moving field, and the emergence of new paradigms such as machine learning present ongoing challenges in keeping teachers—and the curricula they deliver—up to date.

CS in secondary schools is facing an existential threat   Following the good work that has been done to establish CT as a foundational discipline in primary education, one might expect an increase in the number of students opting to study CS in secondary schools. Writing in the *Communications of the ACM* in 2015, Alan Bundy and I said of the future of CT and CS in schools that "it could go either way" (Scott and Bundy 2015, 40). Since then, however, the situation for CS as a discrete subject has become

more precarious across the UK, and the subject faces a number of challenges that could present an existential threat to it in many secondary
schools.

The 2019 Roehampton Annual Computing Education report on the
state of CS education in England revealed that the number of students
electing to take CS in England's secondary schools was stable at around
12 percent of the student population (Kemp and Berry 2019, 3). The
overall number of hours of computing education (including ICT) taught
in English secondary school classrooms, however, fell by 36 percent
between 2012 and 2017 (Kemp and Berry 2019, 2). The report's authors
say that for the majority of students who do not opt to study GCSE computer science (from approximately age fourteen to sixteen), it now looks
unlikely that they will receive any computing education in schools there-
after (BBC News 2019).

The picture is similar elsewhere in the UK, with Scotland facing significant difficulty recently. Analysis of statistics published by the Scottish
Qualifications Authority (Scotland's national awarding body) reveals a
20 percent decline in uptake of Scotland's equivalent National 5 qualification in computing science (studied from age about fifteen to sixteen)
between 2016 and 2019, with numbers in the Higher (Scotland's principal university entrance qualification, taken at around age seventeen to
eighteen) down more than 28 percent in the same period (Scottish Qualifications Authority n.d.).

So, despite everything that has been done in recent years, why is CS in
such decline, and what can be done to turn things around?

Availability of teachers   Arguably, the most pressing issue faced by schools
is the availability of qualified teachers. It is simply not possible to offer a
subject in which you cannot recruit teachers and the number of schools
where the subject is offered for public examination remains a concern. In
England, 54 percent of schools do not offer CS as a discrete subject (The
Royal Society 2017, 45). CAS Scotland reported in 2016 that 17 percent
of Scottish secondary schools did not have a specialist CS teacher and
the number of CS teachers overall had fallen by 25 percent in a decade
(Scotland 2016).[23]

Target and actual recruitment figures to initial teacher education programs in CS indicate recent shortfalls (Robertson 2019), and only 6 percent

of CS graduates have embarked on a career in education compared to 11 percent for physics and 12 percent for mathematics (The Royal Society 2017). In Scotland, the number of first-year students on CS initial teacher training courses fell by 80 percent between 2008 and 2017, with some universities dropping their teaching qualification course in CS (The Royal Society 2017).

While the lack of qualified CS teachers is certainly a worldwide problem—even the US saw only seventy-five new CS teachers graduate in 2016 (Code.org 2017)—this must not be used as an excuse for inaction and effectively giving up.

Recent research gives cause for optimism, however; the situation is not insurmountable and needs only a small proportion of CS graduates to transform the situation. For example, it is believed that Scotland would require only fifty to sixty new secondary school CS teachers per year, against a backdrop of Scotland's universities producing four thousand computing graduates per year (Robertson 2019). Furthermore, almost half of recent computing graduates surveyed gave positive responses about the prospect of teaching CS as a career option.

Teaching is not a career that is immediately visible to students being wooed by employers at careers events, however, so there is a role for university CS departments to highlight teaching as a career option to their CS students. This could be as simple as staging talks by recently qualified CS teachers who could evangelize about the rewards and benefits of a career in teaching to undergraduates who may not have considered it previously.

Other initiatives are having an impact. Teach First is a charity and program that recruits graduates into teaching and is now the second largest graduate recruiter in England and Wales with 25 percent of its graduates from STEM subjects (Teach First 2019).

In Scotland, initiatives to get new teachers into the classroom more quickly for STEM subjects include help for former teachers looking to return; bursaries to ease the transition into initial teacher education for those looking to change career (STEM Bursary Scotland 2019); enabling teachers to work in both primary and secondary; and offering more joint degrees in teaching and specialist subjects. A new "Computing in the Classroom" elective is also being offered to final-year CS undergraduates of some universities and for which students receive course credits. This includes a placement in a school CS department, providing insight into the life of a secondary school CS teacher.

Remuneration   While the potential for high remuneration for CS graduates in industry is often cited as a reason for shortages of CS teachers, research suggests that it is not always the first reason that undergraduates give when explaining why they did not go into teaching—although remuneration remains a significant a consideration for many (Robertson 2019, 21, 25).

A commonly offered suggestion to introduce differential pay—effectively creating a subject-demand-driven job market within teaching—is a thorny issue. Schemes to provide financial incentives for teachers in shortage subjects have been successful in England (Sims and Benhenda 2022), but this is more contentious in Scotland and may struggle to find widespread support.

Equity of access   For CS to become mainstream in schools, it must appeal to every student, and this is clearly not happening. The ongoing problem of low female uptake also continues to blight the subject.

In 2017, in twenty-five local authorities in England, all the CS entries for public examinations came from boys. At GCSE (England's main public examination, taken at age sixteen) only 20 percent of entries were from female students and only 10 percent at A-level (taken at age eighteen), even though girls are higher achievers than boys at GCSE (Education Business 2018). In Scotland, the number of girls taking the Higher public examination (Scotland's principal university entrance qualification, taken at c.seventeen to eighteen) in CS fell by 28 percent between 2016 and 2019 (and by an alarming 65 percent in CS-related subjects between 2009 and 2019). This is against a backdrop of around half of Scotland's registered CS teachers being female, thereby providing female role models in the classroom.[24]

In the UK, CS tends to be chosen by students from more privileged backgrounds. Roehampton report coauthor Peter Kemp recently said, "It looks likely that hundreds of thousands of students, particularly girls and poorer students, will be disenfranchised from a digital education over the next few years" (BBC News 2019).

Tackling this has been in sharp focus at the National Centre for Computing Education, with each local authority assigned a deprivation index so that efforts can be targeted where they are needed most.

In Scotland, the promotion of social inclusion through reducing the attainment gap (between more and less affluent students) and meeting the needs of all learners have been two major drivers in education policy in recent years, with similar indices for social deprivation used to target support.

Have we made CS too difficult?   It has been suggested that some students simply find CS difficult and those looking to maximize their grades will understandably opt for the subjects they feel most confident in. By raising the bar and making CS a rigorous academic discipline, have qualification designers made the subject too difficult, forgetting that public examinations are a game of very high stakes? This is a comment I have heard often when speaking to teachers and academics during my research.

Miles Berry, coauthor of the Roehampton report, said in interview that the new GCSE in computer science has earned a reputation for being more difficult than other subjects, discouraging less academically gifted students:

Even among the academically strong, privileged intake, performance is typically below that of students' other subjects, and thus students, their parents and their head teachers might understandably take the view that this is not an easy way to get top grades. (BBC News 2019).

In Scotland, analysis of SQA statistics (Scottish Qualifications Authority, n.d.) reveals that between 2015 and 2018, CS consistently had one of the lowest A-grade pass rates of any widely offered school subject in major public examinations. If a consensus emerges among students that CS is a difficult option, those seeking to maximize their grades in an increasingly competitive environment may vote with their feet and opt for other subjects. Scotland's new curricular structure has also reduced the number of subjects studied for initial public examinations, creating a squeeze on subjects outside the established core of mathematics, language, and traditional sciences.

It may also be that the current cohort of secondary school students has met CS too late in its school career for foundational knowledge and understanding to have bedded in fully. If so, then the introduction of CT education in primary curricula and fulfilment of CS entitlements may address this issue over the next few years. If, however, the depression of grades and student uptake continues, the relative difficulty of CS qualifications must be addressed.

Subject pathway   The vast majority of CS courses at UK universities do not require applicants to have studied CS at school. This can convey the impression to school students that there is no need to choose CS courses for their public examinations, even among those who may plan to study it at university.

While this is a necessary position for universities to take when not all students have the opportunity to take public examinations in CS at school, it creates a chicken-and-egg situation—both hampering the subject's uptake in schools and requiring universities to start with a lower bar in their first-year courses. It may also help to explain why CS has the highest dropout rate of any subject in UK universities (Higher Education Statistics Agency 2019), as some students may arrive at university without an appreciation of what a CS degree will involve.

A solution to this would be for universities to recommend a school CS qualification for entry to their CS courses (as several already do). As provision in schools improves, universities could then move toward making the qualification an entry requirement, bringing CS in line with many equivalent subjects. This would benefit both schools and universities, with the potential to produce a larger cohort of better-prepared students.

The availability of teaching materials   The scarcest resource for any classroom teacher is time. All too often teachers reinvent the wheel, unaware of excellent existing resources that they could use or adapt with relatively little effort.

Teachers now have access to a huge range of materials, but not all are of high quality, and they are scattered across countless sources. There is a need for curation of materials and alignment with curricula. This is now happening with Barefoot materials in particular as they are mapped to primary curricula across the UK, but there remains an ongoing need for work in this area.

RECOMMENDATIONS

The UK's experiences of introducing computational thinking as a foundational discipline prompt a number of recommendations for those with a similar aim:

RECRUIT AND TRAIN MORE TEACHERS
Universities, government, and industry must work together to address the current shortage of teachers able to deliver CT as well as specialist CS.

SUPPORT TEACHERS
Grassroots organizations like CAS and Barefoot work well, providing support and communication through local hubs where teachers can meet, talk, and share materials, as well as active online communities. Teachers must also be offered high-quality CPD and given the time in which to complete it.

REACH EVERY CHILD
Computational thinking education must be regarded as an entitlement for all learners. It must be embedded into the curriculum from early primary and reach every child from every background. This will, in turn, increase the appeal of CS at the secondary level and help to address the gender imbalance in the subject.

GET PRE-SUBJECT-CHOICE COURSES RIGHT
Prior to choosing subjects for public examinations, students must be exposed to high-quality CS courses that they regard as enjoyable and relevant. The profile of the subject in secondary schools must also be elevated so that choosing to study CS is as natural as choosing chemistry, history, or a foreign language.

GET THE HIGH-STAKES EXAMINATIONS RIGHT
Make courses enjoyable, relevant, and challenging—but don't make them too difficult. The number of passes and top grades for CS in public examinations must mirror those in comparable subjects.

GET THE SUBJECT PATHWAY RIGHT

Universities and schools should coordinate to encourage uptake of CS at school, recommending that prospective CS undergraduates select CS courses for their school public examinations (where they are offered). As provision improves, universities should move toward requiring the subject for entry to CS degrees.

COORDINATION IS VITAL

Universities, learned societies, and industry are taken seriously by the government, so it's important to get their support and to speak together with one voice.

EDUCATE KEY INFLUENCERS

Many politicians, parents, and school leaders have little or no experience of CS themselves. It continues to be important to educate them about what computational thinking is and why it must be a foundational discipline.

THERE'S A ROLE FOR BUSINESS

A significant part of the drive for CT education comes from business, and it is to everyone's benefit that we have a population that can think computationally. Businesses large and small have a role to play here by continuing and expanding their support for CT education at all levels of education.

CONCLUSION

As the leader of a successful school CS department, I remain optimistic.

Our subject is now on a firm intellectual footing, courses have improved significantly, and, importantly, have stabilized after a period of considerable revision. CT is gaining acceptance as an important foundational discipline; we are seeing students arrive at secondary school who have been better prepared in primary; and even the word "algorithm" is now in common parlance!

The subject of CS must also compete in a crowded curriculum—especially against other sciences—so high-quality courses, teaching, and career advice must convince students of its value. Recruitment of specialist teachers remains an issue, but my department recently hosted its first student teachers in several years who, anecdotally, reported that numbers were up significantly in their courses.

Ultimately, it falls to everyone who cares about CT education—teachers, academics, industry figures, and others—to build on the good work that's been done so far. We must continue to work together and speak with one voice. We must be "squeaky wheels" to get that voice heard and hold the attention of policymakers.

Because there is no "them"—there is only us.

ACKNOWLEDGMENTS

Professor Hal Abelson, CSAIL, MIT
Julia Adamson, British Computer Society
Dr Irene Bell, Stranmillis University College
Richard Clement, Cardiff Schools Service
Kate Farrell, University of Edinburgh
William Hardie, The Royal Society of Edinburgh
Martin Hill, George Heriot's School
Professor Siu-Cheung Kong, Education University of Hong Kong
Dale Lane, IBM UK
Brendan McCart, STACS
Judith McColgan, George Heriot's School
Professor Simon Peyton-Jones, Microsoft Research
Jenni Richmond, Methodist College
Dr Judy Robertson, University of Edinburgh
Lucinda Tuttiett, Barefoot Computing Project
Jane Waite, Queen Mary University of London
Magda Wood, Micro:bit Educational Foundation

NOTES

1. At the time of writing, the Northern Ireland Assembly is in a period of suspension, after it collapsed in January 2017 due to policy disagreements between its power-sharing leadership, resulting in little development in education policy during this time.

2. The terms "computer science" and "computing science" are both used across the UK to refer to the same subject. For the purposes of this chapter, the term "computer science" or the abbreviation "CS" will be used—although in Scotland, the "C" stands for "computing," which arguably conveys a stronger sense of study of the discipline and what the technology can be used for, as opposed to the technology itself.

3. Most commonly Microsoft Office.

4. Acorn went on to rebrand as ARM (initially from "Acorn RISC Machines," then "Advanced RISC Machines"), a company that designs the processor chips that power many current portable devices.

5. Most commonly Microsoft Office.

6. As the subject was called then. New courses emerged called "Computing" and "Information Systems" and, since 2015, the course has been called "Computing Science."

7. https://www.computingatschool.org.uk.

8. A term used to describe computational thinking teaching resources that are done without the use of a computer.

9. The UK's largest provider of STEM education and careers support to schools, colleges, and community groups.

10. A charity founded in 2009 to promote the study of computer science in schools. It also developed the Raspberry Pi single-board computer.

11. Peyton Jones described these hubs as CAS hubs' "mother ships" in discussion with the author, 18 September 2019.

12. A modification of Scratch that has since been updated and renamed *Snap!*

13. https://www.compednet.com.

14. https://www.gov.scot/publications/scottish-technology-ecosystem-review.

15. https://www.stacs.scot/home.

16. http://www.cas.scot/plan-c.

17. Richard Clement (Cardiff Schools Service), in discussion with the author, 21 October 2019.

18. Dr Irene Bell (Head of STEM, Stranmillis University College), in discussion with the author, 4 September 2019.

19. Dale Lane (IBM UK), in discussion with the author.

20. https://www.bebras.uk/.

21. https://codeclub.org/en/.

22. https://helloworld.raspberrypi.org/.

23. Cutts et al. cite approximately 640 practicing computing teachers across 420 secondary schools in Scotland (Cutts, et al. 2017, 34), while Robertson's study suggests that this number had fallen to 595 by 2019 (Robertson 2019, 2).

24. A freedom of information request to the General Teaching Council of Scotland (Scotland's professional registration and standards body for teachers) by the author

in 2017 revealed that there were 1,595 registered Computing teachers in Scotland (794 female and 801 male). It should be noted that not all registered teachers will be actively teaching at any time, however.

REFERENCES

Arthur, Charles. 2012. "How the BBC Micro Started a Computing Revolution." Accessed 10 January. https://www.theguardian.com/education/2012/jan/10/bbc-micro -school-computer-revolution.

Barefoot Computing. 2020. "About Barefoot." Accessed October 2020. https://www .barefootcomputing.org/about-barefoot.

BBC News. 2019. "Computing in Schools in 'Steep Decline.'" 7 May. Accessed August 2019. https://www.bbc.co.uk/news/technology-48188877.

Belfast Telegraph. 2016. "NU Education Minister Weir Aims to Get Children into Computers." 07 December. Accessed August 2019. https://www.belfasttelegraph.co .uk/business/news/nu-education-minister-weir-aims-to-get-children-into-computers -35273839.html.

Berry, Miles. 2015. *QuickStart Primary Handbook*. Swindon: British Computer Society.

Code.org. 2017. "Universities Aren't Preparing Enough Computer Science Teachers." 1 September. https://codeorg.medium.com/universities-arent-preparing-enough-compu ter-science-teachers-dd5bc34a79aa.

Computing at School. 2019. "Computing at School: About Us." 5 November. https:// www.computingatschool.org.uk/about.

Council for the Curriculum, Examinations and Assessment. 2018. *Computing at School: Northern Ireland Curriculum Guide for Post Primary Schools*. CCEA.

Cutts, Quintin, Judy Robertson, Peter Donaldson, and Laurie O'Donnell. 2017. "An Evaluation of a Professional Learning Network for Computer Science." *Computer Science Education*. https://doi.org/10.1080/08993408.2017.1315958.

Discovery Research. 2017. "Research | micro:bit." https://microbit.org/impact/re search/.

Donaldson, Margaret. 1978. *Children's Minds*. Harper Perennial.

Education Business. 2018. "BCS Report Reveals Low Uptake of Computer Science at GCSE and A Level." 18 June. http://educationbusinessuk.net/news/18062018/bcs -report-reveals-low-uptake-computer-science-gcse-and-level.

Higher Education Statistics Agency. 2019. "Experimental Statistics: UK Performance Indicators 2017/18." 7 March. https://www.hesa.ac.uk/news/07-03-2019/experimental -uk-performance-indicators.

Independent Schools Council. 2018. "Research—ISC." https://www.isc.co.uk/re search/.

Kemp, Peter E. J., and Miles G. Berry. 2019. *The Roehampton Annual Computing Education Report: Pre-Release Snapshot from 2018*. London: University of Roehampton.

Meredith, Robbie. 2018. "NI Schools See Rise in Pupil Numbers." 13 December. https://www.bbc.co.uk/news/uk-northern-ireland-46538403.

National Centre for Computing Education. 2020. *Impact Report: November 2020*. National Centre for Computing Education.

Papert, Seymour. 1980. *Mindstorms: Children, Computers, and Powerful Ideas*. Basic Books.

Perry, Caroline. 2015. *"Coding in Schools."* Briefing Paper (NIAR 65–15), Northern Ireland Assembly. http://www.niassembly.gov.uk/globalassets/documents/raise/publications/2015/education/3715.pdf.

Robertson, Judy. 2019. *Towards a Sustainable Solution to the Shortage of Computing Teachers in Scotland*. The Centre for Research in Digital Education.

Rousseau, Jean-Jacques. 1762. *Emile, or On Education*. The Hague: Jean Néaulme.

Scotland, Computing at School. 2016. *Computing Science Teachers in Scotland*. Computing at School Scotland.

Scott, Jeremy, and Alan Bundy. 2015. "Creating a New Generation of Computational Thinkers." *CACM* 37–40. https://doi.org/10.1145/2791290.

Scottish Government. 2022. "Preparing Pupils for Careers in Tech." 7 February. https://www.gov.scot/news/preparing-pupils-for-careers-in-tech-1/.

Scottish Qualifications Authority. n.d. "Statistics archive—SQA." Accessed August 31, 2021. https://www.sqa.org.uk/sqa/57518.8313.html.

Sims, Sam, and Asma Benhenda. 2022. *The Effect of Financial Incentives on the Retention of Shortage-Subject Teachers: Evidence from England*. CEPEO Working Paper No. 22–04, University College London, Centre for Education Policy and Equalising Opportunities, UCL. CEPEO Working Paper No. 22–04.

STEM Bursary Scotland. 2019. "STEM Bursary Scotland." Accessed October 2019. https://stembursaryscotland.co.uk/.

Teach First. 2019. "Teacher Recruitment." https://www.teachfirst.org.uk/teacher-recruitment.

Tedre, Matti, Peter J. Denning, and Tapani Toivonen. 2021. "CT 2.0." *21st Koli Calling International Conference on Computing Education Research (Koli Calling '21)*. Joensuu.

The Royal Society. 2017. *After the Reboot*. The Royal Society.

The Royal Society of Edinburgh. 2016. "Resources." Accessed October 2019. https://www.rse.org.uk/schools/resources/.

The Royal Society. 2012. *Shut Down or Restart? The Way Forward for Computing in UK Schools*. The Royal Society.

UK Office for National Statistics. 2011. *UK Census*.

Vasko, Tibor, and Darina Dicheva. 1986. *Educational Policies: An International Overview*. International Institute for Applied Systems Analysis.

Wing, Jeanette M. 2006. "Computational Thinking." *Communications of the ACM (ACM)* 49, no. 3: 33–35.

2

COMPUTATIONAL THINKING EDUCATION IN HONG KONG

Siu-Cheung Kong and Wai-Ying Kwok

INTRODUCTION

The young generation in the twenty-first century is facing a fast-changing digital future, in which the computing-rich society requires citizens to leverage the capacity of digital technology to solve problems in daily life. The competency of computational thinking (CT)—which is a thinking process to systematically identify real-life problems and empathically formulate them for possible computational solutions—has become essential for success in the digital era in the twenty-first century. This entails the need for a school curriculum to engage students—as young as at the primary schooling grades—in CT development. In response to this emerging curriculum need, primary schools around the world have set computational thinking education (CTE) as a top priority in e-learning plans to develop and implement the CT curriculum.

As an important hub of e-learning in the Asia-Pacific region, Hong Kong certainly must account for this educational trend to prepare its youth to be creative problem-solvers in a computing-rich society. Since 2016, a cross-year international initiative has been emerging in Hong Kong primary schools for pioneering a CT curriculum tailored to senior primary students from grades 4 to 6. This initiative is the world's first large-scale CTE pilot that holistically plans a CT curriculum tailored to senior primary grades and consistently supports school implementation of CTE aligning with a school-based e-learning blueprint.

This chapter shares the Hong Kong–based experiences and insights gained from this citywide CTE initiative. The goal of this chapter is to provide researchers and practitioners in K–12 education with inspirations for the planning, development, and implementation of CT curricula favorable for young students in K–12 schools from the Asia-Pacific perspective.

The discussion in this chapter is guided by the Darmstadt Model (Hubwieser 2013; Raman et al. 2015) to dissect the Hong Kong–based policy and curriculum for CTE, taking the relevant educational areas into consideration. A comprehensive picture of CTE in Hong Kong is presented, looking at the educational system, sociocultural-related factors, government policies, readiness of students and teachers, knowledge scope, curriculum issues, examination certification, teaching methods, instructional media, and extracurricular activities for CTE.

POLICY LANDSCAPE FOR COMPUTATIONAL THINKING EDUCATION IN SCHOOL EDUCATION IN HONG KONG

Since 1998, the Hong Kong government has implemented four rounds of the information technology in education (ITEd) strategy to steer the citywide policy on e-learning in basic education in Hong Kong—covering the preschool, primary school, secondary school, and special school sectors. The fourth ITEd strategy—"Realising IT Potential • Unleashing Learning Power"—has been implementing since the 2015/2016 school year (Education Bureau 2015). This latest ITEd strategy sets one of the targets to strengthen students' self-directed learning, problem-solving, collaboration, and computational thinking (CT) competency. This forms the driving force for computational thinking education (CTE) in Hong Kong school education.

For primary school education, the component of CT development is officially set to focus on the key stage 2 student pool (grade 4 to grade 6 students aged around nine to twelve) in all local primary schools and is subsumed into the curriculum initiative of coding education under the strand of STEM (science, technology, engineering, and mathematics) education. The Hong Kong government shapes coding education to be a noncompulsory component in the primary school curriculum. Local primary schools are recommended to embed the component of coding education as a part of the school-based e-learning plan, with flexibility to tailor school-based directions of CT development through coding education according to their own educational needs. As an infrastructural support, the Hong Kong government in the 2017/2018 school year completed the establishment of a Wi-Fi campus for all publicly funded primary schools and disbursed a one-off grant for each school to acquire mobile computing devices for coding education.

The educational authority unit Curriculum Development Council (CDC) Committee on Technology Education, under the CDC at the Education Bureau, takes charge of the e-learning target on CT development through coding education. In November 2017, the Education Bureau released a draft of the strategic document "Computational Thinking—Coding Education: Supplement to the Primary Curriculum" (hereafter referred to as the Supplementary Document) to provide local primary schools with references and recommendations on the planning and implementation of coding education for CT development (Education Bureau 2020). In February 2019, an Ad Hoc Committee for Reviewing the "Computational Thinking—Coding Education: Supplement to the Primary Curriculum" was established for a twenty-month work period to finalize the Supplementary Document, with a thorough discussion about the expected learning contents, lesson time, and implementation modes of coding education in the primary school sector.

The emphasis of CT development in the governmental e-learning blueprint drives local primary schools to prioritize CTE as an area of concern in the school-based development plan. As CTE is a noncompulsory component of the school curriculum, Hong Kong primary school teachers are not provided with compulsory training on CT development through coding education. Meanwhile, the increasing demand of teaching professionals for CTE drives the educational authorities and the tertiary institutions in Hong Kong to offer local primary school teachers the various kinds of noncompulsory yet certificate-awarding training activities in relation to CT development through coding education.

Since the 2017/2018 school year, the Education Bureau of the Hong Kong government has organized five phases of professional development for local in-service teachers. The relevant training is mainly delivered in three ways: (1) foundation courses about the basic theories and concepts of CT, coding, and CT development through coding education; (2) training courses about the pedagogical integration of coding education into STEM-related activities and different subject areas for CT development; and (3) hands-on workshops about the instructional resources for CT development through coding education. Such government-hosted professional development has benefited more than one thousand teachers from more than three hundred primary schools.

Local tertiary institutions also echo the governmental direction in teacher training for CTE. For instance, the authors' affiliated university has been offering a five-week "Certificate in Professional Development Programme on Coding Mobile Apps for Computational Thinking Development," with one batch every April since 2017. This is a government-funded full-time block release program for local in-service primary school teachers to boost the competency and confidence in the delivery of coding education in primary classrooms. This program aims to equip primary school teachers with sound knowledge and hands-on practices of coding for CT development—particularly coding mobile apps, coding for interacting with digital physical objects, and coding for CT development. The intensive training offered in this program has helped around one hundred primary school teachers strengthen their capacity to design and implement pedagogies in the delivery of coding education for CT development in primary classrooms through coding.

The overarching direction of CTE in Hong Kong is to foster students' abilities to creatively tackle complex problems through design and coding. In the Supplementary Document (Education Bureau 2020), the Hong Kong government expects the CT curriculum to prepare senior primary students to be competent in the basic CT concepts and practices; skillful in the process of problem-solving through developing computational programs and processing computational data; and aware of the connection of coding with real-life problems and other subjects.

Hong Kong has no citywide standard on the curriculum structure, content scope, textbook production, or examination certification for the CT curriculum in primary schooling—as there is no formal-subject curriculum officially established for CTE. Instead, the Hong Kong government recommends two areas of concern to the school-based CT curriculum: (1) to cover three basic concepts of CT, namely abstraction, algorithms, and automation; and (2) to link with curriculum subjects for STEM education. Local primary schools are provided with the flexibility to make school-based design and implementation of related components and activities in the CT curriculum based on the Supplementary Document (Education Bureau 2020).

The educational authorities suggest that local primary schools ensure three types of learning opportunities in CT curriculum activities: (1) the opportunity for students to engage in programming activities to develop programming-related capabilities—including the ones of modeling, coding, testing, and analyzing—for CT development during the problem-solving process; (2) the opportunity for students to use various digital technologies in the hands-on and minds-on learning activities for a logical problem-solving process through collaboration and repeated trials; and (3) the opportunity for students to use robots and programming in the general studies and/or the computer-related subjects for nurturing the ability to solve daily-life problems from personal to community levels.

Local primary schools are advised to allocate about ten to fourteen hours for the school-based CT curriculum at each grade across the three-year senior primary schooling. The school-based CT curriculum is recommended to be delivered through either the provision of a school-based program, or the approach of cross-subject theme-based teaching. Extracurricular activities such as intraschool and/or interschool competitions on STEM education are encouraged for students to demonstrate their CT competency outside classrooms. The Hong Kong government recommends local primary schools to adopt student-centered pedagogies in a school-based manner when implementing coding education for CT development.

The next section of this chapter shares a large-scale pilot for pioneering the CT curriculum in Hong Kong primary schools. This cross-year curriculum pilot makes reference to the Supplementary Document for tailoring a CT curriculum in the primary school sector.

CURRICULUM PILOT FOR PIONEERING COMPUTATIONAL THINKING CURRICULUM IN HONG KONG PRIMARY SCHOOLS

In April 2016, an international collaborative initiative titled "Developing Computational Thinking of Senior Primary School Students through Programming Education" launched the citywide curriculum pilot for CTE in the Hong Kong primary school sector (The Hong Kong Jockey Club Charities Trust 2020). This pioneering initiative responded to the governmental policy blueprint for CT development through coding education under the curriculum initiative in STEM education (HKSAR Government News 2018).

The CoolThink@JC initiative embraces a rounded vision of CT to go beyond typical coding education. This initiative's target is to prepare students to be digital problem-solvers who appreciate the value of coding to their own futures and to the whole world (Shear et al. 2020). This large-scale curriculum initiative develops a progressive three-year sequence of lessons that provides CTE based in the visual programming languages Scratch and App Inventor. A cyclical implementation is arranged for the three-level CT curriculum at grade 4 to grade 5 to grade 6, of which level 1 is implemented again for each new cohort of grade 4 students within a school (The Hong Kong Jockey Club Charities Trust 2020). The cross-year curriculum pilot completed its first phase in August 2020, directly benefiting thirty-two local primary schools with more than one hundred teachers and eighteen thousand senior grade students aged around nine to twelve in those primary schools (Shear et al. 2020).

The CoolThink@JC initiative is designed with two intents. First, it aims to develop a CT curriculum that inspires students to apply digital creativity in daily life for problem-solving and nurtures their proactive use of technologies for social good from a young age. Second, it aims to offer a CTE journey that fosters students to transform from passive consumers to active designers and shapers of the digital future (The Hong Kong Jockey Club Charities Trust 2020).

The CoolThink@JC initiative cocreators understand that the level of motivation and interest of students and teachers are important elements in the persistent and successful development of CT competency (Kong, Chiu, and Lai 2018; Kong, Lai, and Sun 2020). Therefore, two emphases are placed in the design rationale of the CoolThink@JC curriculum. First, the target CT curriculum strives to increase students' interest in the learning of and engagement in coding, and so the motivation to develop CT competency for problem-solving in daily life. Second, the target CT curriculum strives to increase teachers' motivation for enhancing CT competency and teaching performance in CT lessons and to widen the vision for CTE emphasizing creative problem-solving on top of coding skills. The CT curriculum developed in the CoolThink@JC initiative covers three knowledge areas with the goal of cultivating students to be creative problem-solvers in the digitalized society.

The first knowledge area is CT concepts, which are the key constructs and ideas central to most forms of computing. Through this knowledge area, students learn about nine concepts designers engage with as they program, including (1) sequences—to identify a series of ordered steps for solving a programming task; (2) events—to understand one thing causing another thing to happen; (3) conditionals—to make decisions based on conditions; (4) operators—to support mathematical and logical expressions; (5) parallelism—to make things happen at the same time; (6) repetition—to run the same sequence multiple times; (7) naming and variables—to store information to be referenced and computed in a program; (8) data structures—to understand basic ways data are stored, retrieved, and updated; and (9) procedures—to create code blocks to modularize and abstract sequences of commands.

The second knowledge area is CT practices, which are the activities people engage in when creating computational projects for problem-solving. Through this knowledge area, students learn about five practices designers develop as they engage with the concepts, including (1) testing and debugging—to make sure things work, otherwise find and solve problems when they arise; (2) being incremental and iterative—to develop a little bit, then try it out, then develop more; (3) reusing and remixing—to make something by building on existing code, projects, or ideas; (4) abstracting and modularizing—to explore connections between the whole and the parts; and (5) algorithmic thinking—to articulate a problem's solution in well-defined rules and steps.

The third knowledge area is CT perspectives, which are the identity and motivation for the daily application of CT for problem-solving in the digitalized society. Through this knowledge area, students learn about five perspectives designers form about the world around them and about themselves, including (1) expressing—to create and express ideas through this new medium; (2) questioning—to feel empowered to ask questions about and with technology; (3) connecting—to appreciate that others are engaged with and appreciate one's creations; (4) digital empowerment—to develop the ability to see problems in the world as solvable through code; and (5) computational identity—to see oneself as being able to enhance the world through coding.

Initiative-specific CT curriculum packages together with a cloud-based learning platform titled "CoolThink@JC Learning Platform" are provided for the participating schools for adoption or adaption in the school-based CT lessons. Following the governmental policy on coding education, no examination is arranged in the CoolThink@JC curriculum. Instead, two citywide extracurricular activities are organized under the CoolThink@ JC initiative annually for students from the participating schools and other local primary schools. First, the "CoolThink@JC Competition" was organized in 2018, 2019, and 2020 to enhance students' creativity, innovation, and collaboration for problem-solving through programming. Second, the "CoolThink@JC Coding Fair" was organized in 2018 and 2019 to promote and facilitate students' CT development through a series of workshops on CT and coding as well as a series of interactive booth exhibitions in the field of CT and STEM.

Apart from curriculum development, the CoolThink@JC initiative also tailors the pedagogical designs for delivering the CT curriculum. In line with the governmental recommendations on teaching approaches in the CT curriculum, the CT curricular activities in the CoolThink@JC initiative are designed to engage students in student-centered learning with a hands-on, minds-on, and joyful learning experience in coding tasks.

The pedagogy "To Play, To Think, To Code, To Reflect, To Create" was designed to support the learning process in the Scratch and App Inventor block-based programming environments (Kong, Lai, and Sun 2020). Students are engaged in "playing" the target apps for inquiry-based learning of the target subject knowledge; then "thinking" about the target subject knowledge through the guided-discovery worksheets; then "coding" in

Scratch and App Inventor for a computational solution in the target problem; then "reflecting" on the target subject knowledge when checking the computational solution in their coding products; and finally "creating" a computational solution for the extension of the target problem through Scratch and App Inventor. Figures 2.1 and 2.2 illustrate CoolThink@JC curricular activities for learning to code with Scratch and App Inventor.

The CoolThink@JC initiative provides teachers in the participating schools with two types of professional development for empowering

(1) To **play** the Scratch app "Maze Game"

(2) To learn CT concepts and practices for app coding

(3) To **think** about worksheet tasks for "Maze Game"

(4) To **code** the Scratch app "Maze Game"

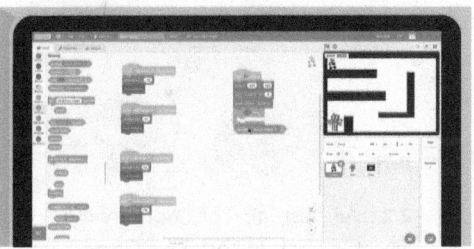

(5) To **reflect** on CT concept "sequence" in app coding

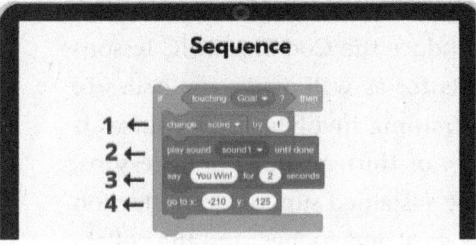

(6) To brainstorm to **create** apps for controlling drone

2.1 An illustration of CoolThink@JC curricular activities for learning to code with Scratch—Unit of "Maze Game."

(1) To **play** the App Inventor app "Addition Game"

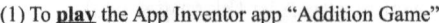

(2) To learn CT concepts and practices for app coding

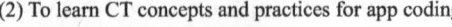
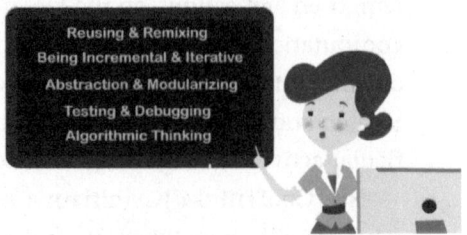

(3) To **think** about the contents for "Addition Game"

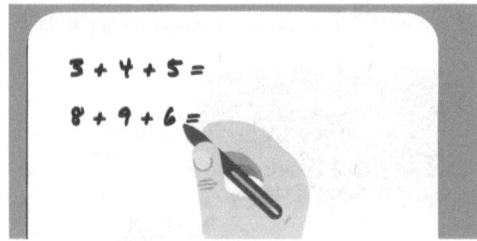

(4) To **code** the App Inventor app "Addition Game"

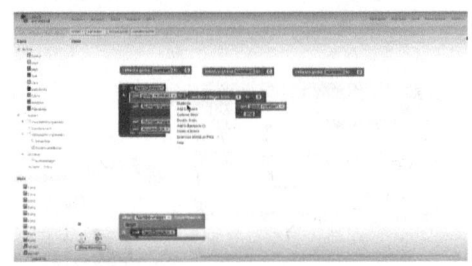

(5) To **reflect** on CT practice "abstracting and modularizing"

(6) To brainstorm to **create** running games

2.2 An illustration of CoolThink@JC curricular activities for learning to code with App Inventor—Unit of "Addition Game."

their capability to deploy the CT curriculum in subject classrooms. First, a one-week workshop is offered to introduce the CoolThink@JC lessons and coding with Scratch and App Inventor as well as discuss issues of pedagogy. Second, a thirty-nine-hour training involving two courses is offered. Each course consists of a series of thirteen three-hour lessons, approximately one per week, to provide sustained support for reflection on instruction, consideration of pedagogical approaches, and the collaborative development of a unit of instruction for CT lessons (Shear et al. 2020; The Hong Kong Jockey Club Charities Trust 2020).

CROSS-YEAR REFLECTION ON PIONEERING COMPUTATIONAL THINKING CURRICULUM IN HONG KONG PRIMARY SCHOOLS

The CoolThink@JC initiative designs and implements a three-level CT curriculum for senior primary students in thirty-two primary schools in Hong Kong—with level 1 for grade 4, level 2 for grade 5, and level 3 for grade 6. Each level of the CT curriculum comprises ten sessions of coding tasks and two sessions of final project. The curriculum pilot is perceived to have a positive impact on fostering students to develop the concepts, practices, and perspectives essential for CT; catalyzing teachers to pedagogically integrate CT elements in subject classrooms for CT development; and inspiring schools to blueprint school-based development of coding education addressing STEM goals—in response to the calls from the Hong Kong government for stepping STEM education up in local primary schools (Shear et al. 2020).

This section reflects on the four-year curriculum pilot in the first phase of the CoolThink@JC initiative. Five lessons learned are discussed for the design and implementation of a CT curriculum desirable for primary school education in the twenty-first century.

LESSON LEARNED (1): THE CT CURRICULUM SHOULD PROVIDE YOUNG STUDENTS WITH AMPLE OPPORTUNITIES TO LEARN FROM PROGRAMMING EXPERIENCE FOR DEVELOPING CT

The introduction of programming education is considered best to cultivate CT competency by engaging students in programming activities for the process of instructing computers to make decisions to solve programming problems (Buitrago Flórez et al. 2017; Guzdial 2019). Computers execute programs—the instructions designed by programmers—to solve problems. Researchers in the relevant fields explicitly link CT to computer programming for a rich multiplicity of approaches to CTE (Guzdial 2019; Shute, Sun, and Asbell-Clarke 2017). Programming education therefore becomes a key component of CTE in K–12 to prepare students to meaningfully use programming to solve problems in the real world (Sung, Ahn, and Black 2017; Tuhkala et al. 2019).

The CoolThink@JC initiative takes these advocacies into consideration for a rationale that the initiative-specific CT curriculum should be designed to engage senior primary students in sufficient learning tasks to gain ample programming experience and so to develop target CT competency.

The CoolThink@JC curriculum is designed to first let students explore the possible solution with programming—in which students have opportunities to brainstorm ideas to tackle the coding tasks to solve a problem. Students are then introduced to the relevant knowledge and/or concepts to facilitate the program development for their work on coding incrementally and iteratively to solve the problem; they then look back at the coding process for a reflection on CT process.

Such a curriculum design is evaluated to be successful to support senior primary students in enhancing the concepts and practices related to coding and extending their programming knowledge and skills in their development of problem-solving and logical thinking skills (Shear et al. 2020). This reveals the practicality of the design rationale behind the CT curriculum that values young students' programming experience for CT development.

LESSON LEARNED (2): THE CT CURRICULUM SHOULD SUPPORT YOUNG STUDENTS ON THE CONNECTION FROM LEARNING SCRATCH TO LEARNING APP INVENTOR

In the CoolThink@JC initiative, the coding tasks in the CT curriculum involve the use of two block-based programming languages—Scratch and App Inventor—in stages. These two block-based programming languages are selected as they are considered favorable for young students in K–12 to use for CTE (Buitrago Flórez et al. 2017; Israel et al. 2015). The drag-and-drop coding method and the hands-on graphical illustration in the Scratch and App Inventor programming environments provide students who have limited or no programming background with a graphically intuitive experience in computer programming (Buitrago Flórez et al. 2017; Kafai and Burke 2014). The CoolThink@JC curriculum pilot therefore selects Scratch and App Inventor to serve as an accessible starting point for young students in grades 4 to 6 to understand how computers follow instructions designed by programmers, and so to use programming to solve problems (Israel et al. 2015; Tuhkala et al. 2019).

From the four-year curriculum pilot, it is observed that in CT lessons with App Inventor starting at grade 5, the students are not observant enough to identify the programming knowledge learned from CT lessons with Scratch at grade 4, and so they are not ready enough to apply the learned programming knowledge for coding in the App Inventor programming environment. Two reasons are possibly behind this observation. First, the code blocks are differently labeled in the two programming environments (see figure 2.3). Second, the user interfaces are differently designed in the two programming environments (see figure 2.4).

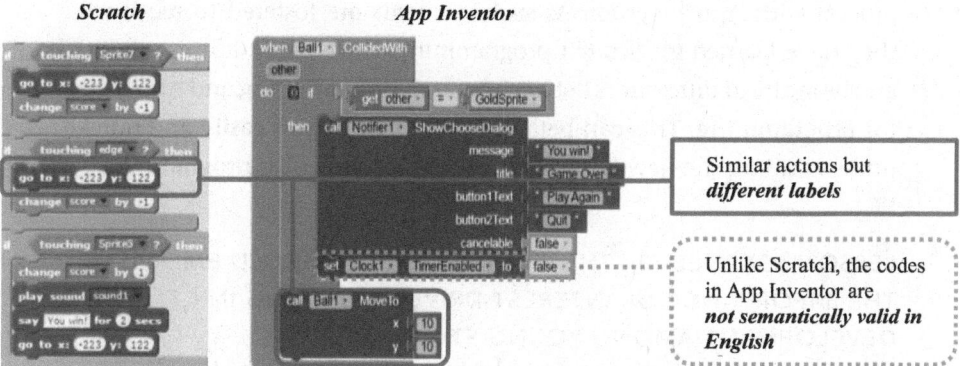

2.3 An illustration showing that the code blocks are differently labeled in the Scratch and App Inventor programming environments.

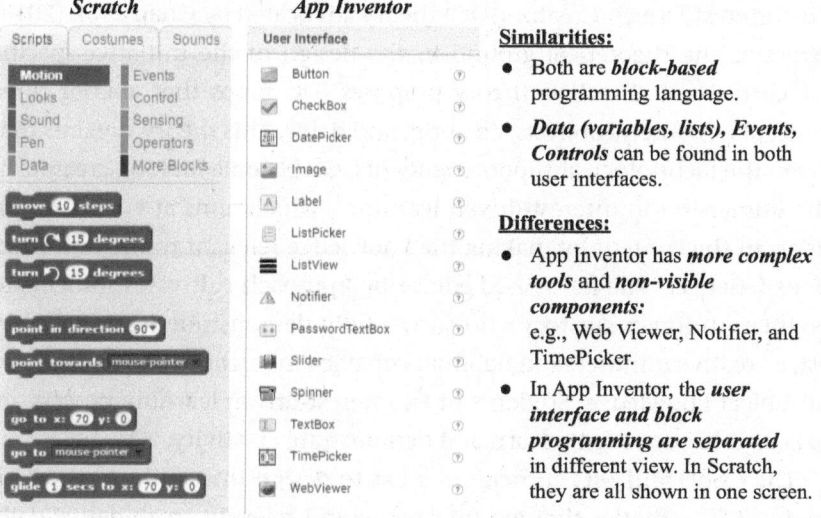

Similarities:

- Both are *block-based* programming language.
- *Data (variables, lists), Events, Controls* can be found in both user interfaces.

Differences:

- App Inventor has *more complex tools* and *non-visible components:* e.g., Web Viewer, Notifier, and TimePicker.
- In App Inventor, the *user interface and block programming are separated* in different view. In Scratch, they are all shown in one screen.

2.4 An illustration showing that the user interfaces are differently designed in the Scratch and App Inventor programming environments.

The previously mentioned experience has driven the CoolThink@JC initiative cocreators to reflect on the corresponding solution in the curriculum design. Students experience a transition from Scratch programming in grade 4 to App Inventor programming in grade 5. During the coding tasks in the App Inventor units in grade 5, teachers are recommended to explicitly guide students to refer to the relevant Scratch programming concepts and practices for the knowledge extension and application in their coding process with App Inventor. As such, students are fostered to recall what they have learned for Scratch programming as well as link up similarities and be aware of differences between Scratch programming and App Inventor programming. This can better support students in easily and quickly understanding the new knowledge about App Inventor programming.

LESSON LEARNED (3): THE CT CURRICULUM SHOULD EMPHASIZE THE IMPORTANCE OF INTEREST-DRIVEN LEARNING IN CT DEVELOPMENT AMONG YOUNG STUDENTS

The CoolThink@JC initiative cocreators recognize well that interest is the ideal motivation for learning. One key principle in designing the initiative-specific CT curriculum is therefore to "use interest-driven activity design as a strategy to incubate interest-driven creator" (Kong 2016). The Interest-Driven Creator (IDC) theory advocated by Chan et al. (2018) serves as the theoretical ground in the design of the initiative-specific CT curriculum. The IDC theory proposes IDC loops that anchor three concepts—namely interests, creation, and habit. This theory suggests that given the technological support, students can become lifelong creators in the immersion in interest-driven learning—which aims at engaging students in the content by making the knowledge relevant to their interests or experiences. The IDC-based education approach cultivates students to be interest-driven creators who successfully demonstrate learning interest, a creative mindset, and habitual behavior in learning and application of subject knowledge. Students in an interest-driven learning process are expected to persistently learn and demonstrate creativity.

The CoolThink@JC curriculum seeks to design interesting and meaningful CTE activities that are up to date and relevant to students' daily life for nurturing creators in the digitalized society. In accordance with the previously mentioned rationale of providing students with ample programming experiences for CT development, the CTE activities central in the CoolThink@JC curriculum are in the type of mobile apps programming. The two selected programming environments—Scratch and App Inventor—in the CT curriculum serve as a good motivation for the interest-driven learning of programming knowledge and CT competency, as these block-based programming contexts enable novice programmers to easily and creatively design simple and complicated games, animation, and mobile apps.

It is important for IDC curriculum designers to understand what students regard as interesting and meaningful in the target subject areas in order to design motivating learning activities and create relaxing learning environments for both students and teachers (Chan et al. 2018). The CoolThink@JC initiative cocreators therefore conducted a survey (Kong 2017) to understand what mobile apps the grade 4 to grade 6 students in Hong Kong are interested in. Games are found to be the most popular apps among local senior primary students. Other functional apps for the purposes of music enjoyment and social networking are also popular among the young students. In this regard, the three-year CT curriculum is designed to engage students in the coding of simple apps in stages, covering the range of leisure games, educational games, music players, and so on (see figure 2.5).

Such a student-interest-driven approach is confirmed to be important in the pedagogical designs of CTE activities (Kong, Chiu, and Lai 2018). When students' interest in programming is stimulated, they find that it is more meaningful to enhance their programming knowledge and more impactful to use computational solutions for problem-solving in daily life; and they are more ready to apply their programming knowledge for tackling daily problems. This reveals the need for CT curriculum planners to design interesting and meaningful CTE activities for triggering students' interest in programming and so engaging them in CT development.

LESSON LEARNED (4): THE CT CURRICULUM SHOULD ADDRESS THE DIFFICULTY FACED BY YOUNG STUDENTS IN CREATING THEIR OWN CODING PROJECTS

The CoolThink@JC curriculum arranges two sessions of final project per year for providing students with regular opportunities to keep growing

Maze Game
(Leisure game app)

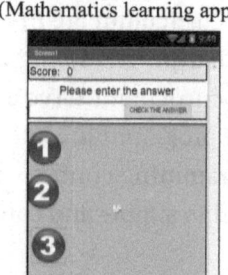

Addition Game
(Mathematics learning app)

Interactive English Quiz
(English learning app)

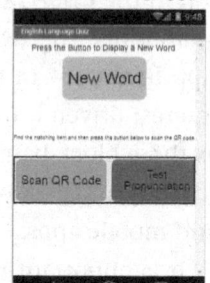

The Three Little Pigs
(Storytelling app)

My Piano App
(Music player app)

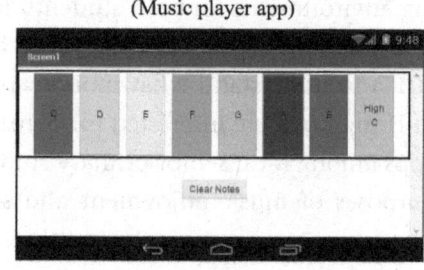

2.5 Examples of mobile apps for CTE under the CoolThink@JC curriculum pilot.

their interest, engagement, and habit in developing coding competency and creating coding products (Kong 2016). Following the ten-session coding tasks that serve to trigger students' programming interest, a two-session final project comes at the end of the CT curriculum each year and serves to maintain students' interest and nurture their creativity in line with the rationale of the creations loop in IDC theory (Chan et al. 2018) that creation consists of imitating, combining, and staging.

Senior primary students are novice programmers; most of them are new to creating, managing, and staging coding projects on their own. The final project guides students through the steps of imitating examples of apps development in the CT curriculum, combining the coding examples with their own ideas, creating their own new mobile apps, and staging their creations of coding products. There are three features supporting the arrangement of the final project. First, the units of the CT curriculum across three years are interrelated. Second, the requirement of completing the final project covers the knowledge learned in the previous CT curriculum units. Third, the students are guided by teachers to apply the programming knowledge and skills learned for finishing the final project (see figures 2.6 and 2.7).

Both the teachers and the students in the CT curriculum pilot find the arrangement of a year-end final project inspiring for the development of digital creativity (Shear et al. 2020). The final project gives students the opportunity to explore their creativity. Students enjoy a lot of autonomy to computationally solve problems by interest, giving them a great sense

LEVEL 3 MIT App Inventor Programming II	Unit 1–2 Hong Kong Tour Guide	Unit 3 Two-button Game	Unit 4–5 Factor Game		Unit 6–8 Sketch & Guess Game			Level 3 Project
LEVEL 2 MIT App Inventor Programming I	Unit 1 Hello It's Me	Unit 2 My Piano App	Unit 3 Music Jukebox	Unit 4 Addition Game	Unit 5–6 Vocabulary Learning App		Unit 7–8 Find the Gold	Level 2 Project
LEVEL 1 Scratch Programming	Unit 1 Creative Computing	Unit 2 Dancing Cat	Unit 3 Making a Maze Game with Scratch	Unit 4 Tell a Joke with Scratch	Unit 5 Tell a Story with Scratch	Unit 6 Make a Magic	Unit 7–8 Computational Arts	Level 1 Project

2.6 The three-level CT curriculum with interrelated units.

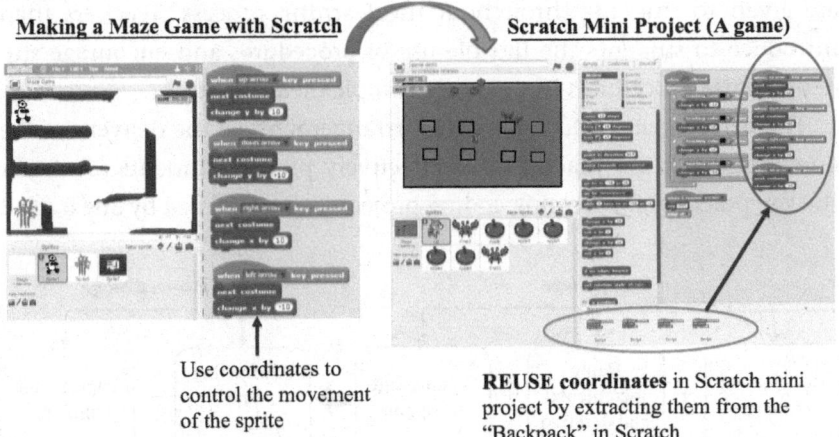

Making a Maze Game with Scratch → **Scratch Mini Project (A game)**

Use coordinates to control the movement of the sprite

REUSE coordinates in Scratch mini project by extracting them from the "Backpack" in Scratch

2.7 An example of a Scratch project that makes use of the knowledge taught in the previous CT curriculum units.

of achievement and satisfaction when they finish the coding products in the final project. This reveals the need for CT curriculum planners to make strategic arrangements such as including a year-end final project for students to review and consolidate CT competency through real-life application.

LESSON LEARNED (5): THE CT CURRICULUM SHOULD PROVIDE YOUNG STUDENTS WITH EXPLICIT GUIDES TO DEVELOP CT PRACTICES IN STAGES

As introduced in the preceding section, the CoolThink@JC curriculum covers the knowledge area CT practices with an emphasis on five practices, namely (1) testing and debugging, (2) being incremental and iterative, (3) reusing and remixing, (4) abstracting and modularizing, and (5) algorithmic thinking. These five practices are delivered in two stages in the CT curriculum, with the initial stage focusing on the first three CT practices and the advanced stage on the remaining two (see figure 2.8).

The delivery of these five practices is featured with the explicit guides from teachers to support students to progressively develop CT practices (see figure 2.9 for an example). Teachers explain explicitly to students the reason behind learning the procedures concerned and the way to relate those procedures with abstracting and modularizing. Timely reminders are given to students throughout the learning process. Teachers then introduce to students the flexible use of procedures and encourage students to use procedures in the more complex learning tasks.

The beforementioned pedagogical arrangements in the delivery of the knowledge area CT practices can effectively prepare students to master the key practices for creating coding projects. As confirmed by Shear et al.

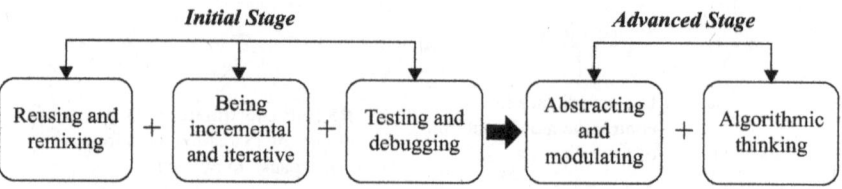

2.8 The flow of pedagogical delivery of the knowledge area CT practices in the CoolThink@JC curriculum.

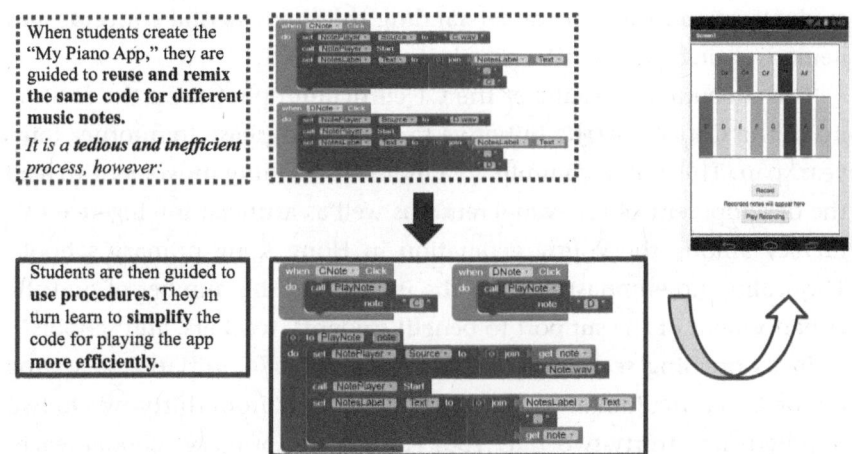

2.9 A coding task for a progression from the practice of "reusing and remixing" to the practice of "abstracting and modularizing."

(2020), students in the CoolThink@JC initiative have a particularly strong achievement in demonstrating CT practices, thereby building an encouraging foundation for developing transferable skills in problem-solving and logical thinking for their learning in other subject disciplines such as mathematics. Such pedagogical success reveals the need for CT curriculum planners to provide young students with a progressive development of CT practices in stages during their CTE process.

CONCLUSION AND SCALING-UP DIRECTIONS

This chapter shares the experiences and insights gained from a cross-year international initiative in a large-scale CT curriculum pilot in Hong Kong, which benefits more than one hundred teachers and eighteen thousand senior primary students across thirty-two local primary schools. A three-level CT curriculum is tailored to grades 4 to 6 students in Hong Kong for nurturing their knowledge of concepts in computer programming and CT; their practices of problem-solving and logical thinking; and their perspectives to get interested and motivated in CT development and application. The design and implementation of the CT curriculum has positive attainments in bringing CTE to local primary schools. The senior primary students perform well in understanding the fundamental concepts and demonstrating the essential practices related to CT.

The encouraging results of the CT curriculum pilot drive the scaling-up of the CoolThink@JC initiative to its second phase in another four-year span. This phase commits to a more far-reaching movement to lead the development of CT competency as well as artificial intelligence (AI) literacy among the young generation in Hong Kong primary schools. The scaling-up emphasizes both the increase in the number of and the enhancement of the support to benefit students, teachers, and schools.

In its ongoing second phase, the CoolThink@JC initiative is scaling up the CT curriculum at the pilot school network from thirty-two to two hundred (i.e., thirty-two plus 168) Hong Kong primary schools reaching the 2023/2024 school year. At the time of writing this chapter, more than 150 local primary schools are joining the CoolThink@JC network for the CT curriculum pilot; more than 280 teachers are being trained for implementing the initiative-specific CT curriculum; and more than twenty thousand students are being taught with the initiative-specific CT curriculum materials.

While keeping the established goal of promoting grade 4 to grade 6 students to develop CT and digital creativity through programming, the initiative is scaled up to extend its connections to the empowerment of "future-ready" students who are competent in technological use and even technological innovation to navigate changes and overcome challenges in their daily lives in the digitalized society. Accordingly, the design of the initiative-specific CT curriculum is refreshed to lead a comprehensive CTE journey of understanding Scratch programming, understanding App Inventor programming; building AI awareness, and exploring programmable robotics with design thinking. Such a scaling-up effort is expected to enrich students' experience in CT curriculum learning, to enhance teachers' readiness for CT curriculum delivery, and to widen schools' vision of a CT curriculum blueprint emphasizing creative problem-solving on top of coding competency. Research supports are planned for the system-level sustainability of the citywide curriculum initiative.

REFERENCES

Buitrago Flórez, Francisco, Rubby Casallas, Marcela Hernández, Alejandro Reyes, Silvia Restrepo, and Giovanna Danies. 2017. "Changing a Generation's Way of

Thinking: Teaching Computational Thinking through Programming." *Review of Educational Research* 87, no. 4: 834–860.

Chan, Tak-Wai, Chee-Kit Looi, Wenli Chen, Lung-Hsiang Wong, Ben Chang, Calvin C. Y. Liao, Hercy Cheng, et al. 2018. "Interest-Driven Creator Theory: Towards a Theory of Learning Design for Asia in the Twenty-First Century." *Journal of Computers in Education* 5, no. 4: 435–461.

Education Bureau. 2015. *Realising IT Potential Unleashing Learning Power: A Holistic Approach*. Hong Kong: Education Bureau. https://www.edb.gov.hk/attachment/en /edu-system/primary-secondary/applicable-to-primary-secondary/it-in-edu/ITE4 _report_ENG.pdf.

Education Bureau. 2020. *Computational Thinking—Coding Education: Supplement to the Primary Curriculum*. Hong Kong: Education Bureau. https://www.edb.gov.hk /attachment/en/curriculum-development/renewal/CT/CT%20Supplement%20 Eng%20_2020.pdf.

Guzdial, Mark. 2019. "Computing for Other Disciplines." In *The Cambridge Handbook of Computing Education Research*, edited by Sally A. Fincher, and Anthony V. Robins, 584–605. Cambridge: Cambridge University Press.

HKSAR Government News. 2018. "LCQ22: Promotion of Education on Subjects Related to Science, Technology, Engineering and Mathematics." Accessed September 24, 2020. https://www.info.gov.hk/gia/general/201811/28/P2018112800583.htm.

Hubwieser, Peter. 2013. "The Darmstadt Model: A First Step towards a Research Framework for Computer Science Education in Schools." In *Informatics in Schools. Sustainable Informatics Education for Pupils of All Ages*, edited by Ira Diethelm and Roland T. Mittermeir, 1–14. Berlin Heidelberg: Springer.

Israel, Maya, Jamie N. Pearson, Tanya Tapia, Quentin M. Wherfel, and George Reese. 2015. "Supporting All Learners in School-Wide Computational Thinking: A Cross-Case Qualitative Analysis." *Computers and Education* 82: 263–279.

Kafai, Yasmin B., and Quinn Burke. 2014. *Connected Code: Why Children Need to Learn Programming*. Cambridge, MA: MIT Press.

Kong, Siu-Cheung. 2016. "A Framework of Curriculum Design for Computational Thinking Development in K-12 Education." *Journal of Computers in Education* 3, no. 4: 377–394.

Kong, Siu-Cheung. 2017. "A Survey Study for Understanding Interest of Primary School Learners in Programming Mobile Apps: Designing an Interest-Driven Curriculum." In *Conference Proceedings of the 21st Global Chinese Conference on Computers in Education 2017*, edited by Maiga Chang, Morris Jong, Tak-Wai Chan, Shengquan Yu, Fati Wu, Baoping Li, Wenli Chen et al., 783–786. Beijing: Beijing Normal University.

Kong, Siu-Cheung, Ming Ming Chiu, and Ming Lai. 2018. "A Study of Primary School Students' Interest, Collaboration Attitude, and Programming Empowerment in Computational Thinking Education." *Computers and Education* 127: 178–189.

Kong, Siu-Cheung, Ming Lai, and Daner Sun. 2020. "Teacher Development in Computational Thinking: Design and Learning Outcomes of Programming Concepts, Practices and Pedagogy." *Computers and Education* 151: 103872.

Raman, Raghu, Smrithi Venkatasubramanian, Krishnashree Achuthan, and Prema Nedungadi. 2015. "Computer Science (CS) Education in Indian Schools: Situation Analysis Using Darmstadt Model." *ACM Transactions on Computing Education* 15, no. 2: 1–36 DOI: http://dx.doi.org/10.1145/2716325.

Shear, Linda, Haiwen Wang, Carol Tate, Satabdi Basu, and Katrina Laguarda. 2020. *CoolThink@JC Pilot Evaluation: Endline Report*. Menlo Park, CA: SRI International.

Shute, Valerie J., Chen Sun, and Jodi Asbell-Clarke. 2017. "Demystifying Computational Thinking." *Educational Research Review* 22: 142–158.

Sung, Woonhee, Junghyun Ahn, and John B. Black. 2017. "Introducing Computational Thinking to Young Learners: Practicing Computational Perspectives through Embodiment in Mathematics Education." *Technology, Knowledge and Learning* 22, no. 3: 443–463.

The Hong Kong Jockey Club Charities Trust. 2020. "CoolThink@JC." Accessed September 24, 2020. https://www.coolthink.hk/en/ct/.

Tuhkala, Ari, Marie-Louise Wagner, Ole Sejer Iversen, and Tommi Kärkkäinen. 2019. "Technology Comprehension—Combining Computing, Design, and Societal Reflection as a National Subject." *International Journal of Child-Computer Interaction* 20: 54–63.

3

A CORE-COMPETENCY-ORIENTED COMPUTATIONAL THINKING EDUCATION IN CHINA

Xiaozhe Yang and Youqun Ren

THE IMPORTANCE OF COMPUTATIONAL THINKING AT THE K–12 STAGE

After moving into the twenty-first century, information technology has been constantly changing with the times, and information society has undergone fast-paced development. While shifting the center from computers and the Internet to data centers, people's work and lives have met with different changes (Ravenscroft 2001). In the meantime, due to the advancement of artificial intelligence of a new generation, information technology has enhanced social productivity and production relations differently and is transforming students' everyday lives, learning styles, thinking patterns, values, and viewpoints (Nouri et al. 2019). Contrary to such a temporal background, the education on information technology is infused with different missions and implications. Reinforcing education on information education and developing students' survival skills and significant personalities in an information society have been accepted as critical issues in the context of existing primary education (Mannila et al. 2014).

Papert (1990) was the first to put forward computational thinking (CT). Subsequently, its definition, teaching, and assessment have been extensively used. Plenty of scholars have analyzed CT (Grover and Pea

2013). Wing (2008) stresses that CT should be among everyone's daily life skills rather than being limited to the programming skills that computer scientists exercise exclusively. The role played by the social development of courses is manifested in the cultivation of digital citizens and in students' awareness of the rights, obligations, and opportunities while they live, study, and work in an interconnected, digitized world so that they can take measures and act in a safe, legal, and sensible way.

Many nations around the world have turned their attention to the teaching of computational thinking at the stage of primary education. In October 2016, the directing committee of the K–12 Computer Science Framework in the US released the "K-12 Computer Science Framework" (abbreviated as the American K–12 CS Framework) (Love and Strimel 2016). As computers are made an integral part of the world, there is an increasing demand for computer science education. At present, plenty of students apply computer science to advance their future professional careers. Nevertheless, the opportunities to study computer science fail to satisfy the demands of the public. Despite access to computers by more and more students on campus, not everyone can have the chance to study computer science. The objective in designing the "K–12 Computer Science Framework" is to present all states, areas, and organizations with requisite construction elements in the development of their own computer science and facilitate the enforcement of construction means, promoting professional advancement as a foundation.

In 2011, the International Society for Technology in Education (ISTE) entered into cooperation with the Computer Science Teachers Association (CSTA) in the provision of an operational definition of the expressive characteristics of computational thinking: "Computational thinking refers to a process of resolving problems that involves the six elements of describing the problem, analyzing data, abstraction, designing algorithms, assessing the optimal proposals and working out solutions" (Grover and Pea 2013).

In September 2013, the Department for Education in the UK published "Computing Program of Study: Key Stages1–4." The existing ICT courses developed in the UK were met with both doubt and criticism (Brown et al. 2014). Many middle school (K6–K9) and primary school (K1–K5) students in the UK felt discontent with the original ICT courses and consider them inflexible and boring. The experts in information technology take the view that British middle schools and primary schools need to teach more serious computer science courses. The relevant courses to computer education are purposed to allow students to gain understanding of and make changes to the world by cultivating computational thinking and creativity. Computer science is put at the center of computing, inclusive of information and computer principles, number systems, and programming. Students should learn how to use information technology for the development of programs and systems according to their personal knowledge and understanding of computer science. In addition, computer education can ensure that students master digital literacy and applyICT to express their own ideas, which enables them to develop adaptivity to future work and play an active role in the digital society. This reflects the essence of ICT courses. Back in 2012, the British Computing at School Working Group stated in its research report that computational thinking represents the process of employing computing tools and technologies to comprehend artificial and natural information systems. It fully manifests logical ability, algorithmic capabilities, recursive facility, and the power of abstraction (Brown et al. 2014).

The Australian Curriculum, Assessment and Reporting Authority (ACARA) released a report in 2014 specifying that digital and technological courses need to serve as a discipline that instructs students on how to apply computational thinking and information systems to address problems (Falkner, Vivian, and Falkner 2014). In 2011, the Department of Curriculum Planning and Development, Ministry of Education, Singapore published a "Computer Application Syllabus," requiring that computer application be a mandatory discipline for all students. The courses in primary schools are all-inclusive disciplines, and all other courses taught in middle schools and high schools (K10–K12) are independent disciplines (Chai, Koh, and Tsai 2013). The courses taught at the primary stage focus on the cultivation of fundamental ICT skills in students to provide support for their study of other disciplines and make a difference to their personal lives. The courses in high schools are required to lay emphasis on advanced thinking skills and concentrate on resolving problems and comprehending basic programming concepts and skills. This is conducive to students' learning of programming and prepares them for future careers.

It is understood that such nations as the UK and the US have offered the technology-oriented information technology courses and have begun to proactively advance the relevant programs and plans so as to incorporate "computer science" into the curriculum of middle schools and primary schools. The development of number literacy and computational thinking capability has grown into an emerging trend of information technology education across not only middle schools but also primary schools (Heintz, Mannila, and Färnqvist 2016).

COMPUTATIONAL THINKING EDUCATION IN CHINA

The progress made in information technology education across China is an integral part of the development of information technology education across the globe. From the end of the 1970s, as computer technology and the Internet developed, information technology education has undergone three phases: early computer education, computer application education, and information technology education. Every single stage has its unique characteristics. These crucial points reflect not only the characteristics of technological development in different historical stages but also the different value orientations adopted in information technology education. During the late 1970s, the emergence of personal computers attracted attention from across the international education community. Many countries around the world incorporated computers into primary and secondary education. In 1981, at the Third World Conference on Computer Education Applications (WCCE/81) in Lausanne, Switzerland, the prominent scholar Yelshoff proposed the idea that human beings live in a "programming world." This view reflects the idea of "computer culture" and lays emphasis on the cultivation of computer culture by delivering programming education.

Subsequently, programming started to be recognized as a significant part of computer teaching. To carry out international education reform, China launched the "BASIC Programming Language" course in five priority universities, including Tsinghua University, Peking University, Beijing Normal University, East China Normal University, and Fudan University in 1982. In addition, China started to practice computer teaching in primary and secondary schools, which is viewed as the prelude to the test.

With the continuous progress made in teaching experiments, the Ministry of Education held two computer education work conferences in primary and secondary schools in 1983 and 1984, respectively, to summarize and reflect on the experimental work. In the meantime, the working principles and syllabus were determined for computer education in primary and secondary schools.

What is most worth noting is that in 1984, when conducting an inspection of children's computer activities in the China Welfare Association Children's Palace in Shanghai, Deng Xiaoping proposed that "the popularity of computers should start from the children." This sentence has turned out to be a historical milestone in the development of computer education for China. Not only does it establish the significant position of the computing curriculum in primary and secondary schools but it also transformed computer teaching from experimental trials to large-scale promotion across the country.

In the years after the mid-1980s, as computer application software increasingly matured, the scope of computer applications expanded rapidly, which prompted many non-computer professionals to start employing these tools in work. At the Fourth World Conference on Computer Education Applications (WCCE/85) held in Virginia in 1985, some scholars proposed an educational goal of using computers as a tool for improved efficiency, with the emphasis placed on the application of computers as a tool to resolve practical problems.

At this time, the focus of computer education had shifted from programming to computer application education based on operational skills training. Under the influence of this "computer tool theory," the former State Education Commission made a decision to increase computer application software education content (such as databases, spreadsheets, and so on) at the "National Secondary School Computer Education Third Work Conference" held in 1986. Premised on the original syllabus, the "Syllabus for the Elective Course of Electronic Computers in Ordinary Middle Schools (Trial)" appropriately reduced the requirements for the programming skills section and incorporated some educational contents for the introduction of computer application software in 1987.

In the same year, to implement the instructions issued by Comrade Deng Xiaoping that "education needs to face modernization, face the world, and face the future," the National Middle School Computer Education Research Center was officially established, while the Beijing Research Department and Shanghai Research Department were established respectively at Beijing Normal University and East China Normal University. As computer education was carried out on a large scale in primary schools, it was subsequently renamed the "National Primary and Secondary School Computer Education Research Center." The establishment of the center has promoted the implementation of information technology education across primary and secondary schools in China while laying a solid foundation for the development of the follow-up curriculum guidelines.

In the late 1980s, China started informatization measurement, assessing the potential development force of the information industry by the number of college students per one hundred people and the proportion of scientific and technological personnel, and so on. In the later rounds of informatization composite index evaluation, the level of information subject and information talents have been paid increasing attention. Since the 1990s, the development of network technology has led to the rapid growth of global information, which makes the effective acquisition and application of information an important capability for social survival. The abundant information resources provide a strong support for people to solve the problems they encounter. In the meantime, however, this also imposes more psychological burden on people and makes their work more stressful. As the basic qualification required for citizens living in the information society, the educational goal of information literacy training has been gradually established. In October 2000, the Ministry of Education held the "National Primary and Secondary School Information Technology Education Work Conference," propagating the "Guidelines for the Information Technology Curriculum for Primary and Secondary Schools (Trial)," "Notice on the Popularization of Information Technology in Primary and Secondary Schools," and "About Primary and Secondary Schools." The important guidance documents such as the "Notice on Implementing the School-School Project" clearly state the necessity to "train students' good information literacy."

In the meeting, it was also decided to use the information from five to ten years in the national primary and secondary schools (including secondary vocational and technical schools) to publicize information from

2001 onward. Since then, the name "Computer Education in Primary and Secondary Schools" has been officially changed to "Information Technology in Primary and Secondary Schools." Technical education and information technology courses are listed as compulsory courses for primary and secondary schools. This conference marks the start of the basic education curriculum in China to embrace a broad, forward-looking vision and the proactive response made by the nation to the fast-paced development of information technology. Information technology education has since been put on a fast track of development.

In 2003, the Ministry of Education promulgated the "Standards for Senior High School Technical Courses (Experiment)" including general technology and information technology, hereinafter referred to as the course standard (experiment). The course standard (experiment) explains the basic concept, course objectives, and course content of the high school information technology course. Its introduction facilitates the development of information technology education in primary and secondary schools in China.

In China, the construction of education informatization has also achieved remarkable progress in the field of basic-level education, as reflected in the significant improvement made to teaching conditions for the development of information technology courses. Since 2000, China has enforced an educational system of information technology where information technology courses are regarded as the primary courses. With the launch of engineering projects such as the "School-to-School Link" project, "Modern Remote Educational Project in Rural Middle Schools and Primary Schools," the information-based facilities in middle schools and primary schools have consistently improved to the extent that the needs of information technology education can be basically met on the whole.

According to the statistics released by the Ministry of Education, by the end of the fourth quarter of 2017, 92.1 percent of the primary and secondary schools nationwide (except for teaching points) had gained access to the Internet, 3.03 million classrooms had been equipped with multimedia teaching equipment, and 86.7 percent of the primary and secondary schools had already been complete with multimedia classrooms. Among them, 62.2 percent of the schools had achieved the full coverage of multimedia teaching equipment. By 2015, 100 percent of high schools,

95 percent of junior high schools, and 50 percent of primary schools offered compulsory courses in relation to information technology. These statistics show that the information technology curriculum implemented in schools has been popularized nationwide.

With respect to teachers, there has been a consistent increase in the overall number of teachers teaching lessons on information technology. The teacher–student ratio has been in decline on an annual basis, and the gap between urban and rural areas has been gradually closed. Overall, China has 108,000 full-time teachers of information technology in primary schools, 88,000 in middle schools, and 38,000 in high schools. In 2012, as revealed by a survey conducted on the implementation of information technology curriculum criteria published by the Ministry of Education, up to 90.9 percent of high school information technology teachers had attained bachelor's degrees or above in the relevant disciplines to computer science, educational technology, and mathematics. A large proportion of high school students had studied information technology courses in primary schools and middle schools, among whom up to 44.3 percent of students believed that the overlapping rate between "information technology basics" and the content they had learned in middle schools ranged between 20 and 50 percent. Though high school students who were not beginners accounted for a large proportion, their understanding of information technology courses remained operational. From a hardware perspective, the conditions of information technology education have improved significantly. Moreover, the overall qualities and capabilities of the teaching and research group have been enhanced, thus laying a solid foundation for the further development of computational thinking education on information technology across both middle schools and primary schools.

Computational thinking has been officially recognized as one of the critical qualities of information technology courses in the curriculum standards of Chinese high schools (Wang 2019). Computational thinking refers to a series of thinking activities performed in the process of coming up with solutions by individuals who apply the means of thinking in the field of computer science. Students capable of computational thinking can define problems, abstract features, construct structural models, and organize data in a reasonable way by using computers to perform information activities. They can work out solutions by making judgments, conducting analyses, consolidating various information resources, and applying suitable algorithms. Moreover, they are able to summarize the process and methods used to solve problems with the assistance of computers and extend them to solving other relevant problems.

CHANNELS FOR PRACTICING COMPUTATIONAL THINKING EDUCATION

Computational thinking represents a core element of computer science. It is interconnected with various layers of K–12 learning, with the emphasis placed on abstraction, automation, and analysis (Lye and Koh 2014). As required by the curriculum standard, students shall acquire the following skills: (1) developing and executing algorithms including sequence, cycle, and conditions to accomplish tasks with or without the assistance of computing equipment; (2) analyzing and debugging algorithms of sequence, events, cycle, conditions, concurrence, and variables; and (3) developing models for the working principles of computer systems. In addition, courses in social development include the influence of computation. A knowledgeable and responsible person should understand the social influence of the digital world, including fairness and the exercise of computation.

Computational thinking refers to a series of thinking activities performed by individuals coming up with solutions through thinking from the perspective of computer science (Brennan and Resnick 2012). It is an explicit skill shown in abstracting the problems and automating solutions, consisting of concrete components and practices that are observable and measurable. It is also a term used to describe a growing focus on students' knowledge development, design computational problem-solving, algorithmic thinking, and coding. The core concepts of computational thinking include algorithmic thinking, abstraction, decomposition, and generalization.

The four characteristics (formalization, modeling, automatic, and systematic) of computational thinking are defined in the China information technology curriculum standards released in 2017. Formalization means that computers can be applied in the process of information activities to define problems, identify crucial elements, and conduct analyses of the relationship between different elements by following the way a computer operates. Modeling refers to the construction of models of information processing, the organization of data in a sensible way, and the ability to propose solutions to problems through judgment, analysis, and a combination of various information resources. Automatic refers to the process and means of resolving problems by applying information technology and realizing automation in the solutions to those problems. Systematic refers to the systematic process of developing solutions and transferring them to the solving of other relevant problems.

Based on the requirements for the cultivation of comprehensive information literacy, unit teaching design is worth emphasizing. A unit refers to a learning unit that revolves around knowledge, skills, comprehensive themes, and project activities based on curriculum standards in the selection of learning materials for the structural organization. The teaching of information technology units focuses on the critical concepts of the discipline, encouraging students to become flexible technology users who are not only capable of understanding and developing information technology with computational thinking but also responsible as the users of information. As for the improvement of information literacy among students, the focus should be placed on the expression of core concepts and the overall conception of knowledge systems. The curriculum needs to revolve around the problem-solving process and give consideration to the expression of information consciousness, computational thinking, digital learning, and innovation in unit teaching. In addition, it needs to reflect on the expression of information security consciousness, information using standards, and information morality principles in unit teaching from the perspective of information social responsibilities.

The core concepts need to be put at the center of design for unit learning activities, problem-solving ought to be regarded as the focus, and the learning requirements of putting it to practice need to be clarified to demonstrate what one has acquired and learning to reason, which is because students' understanding as to the way of thinking is embedded in science and technology. Besides, the threads of technological developments start with the understanding of the significant characteristics of core concepts, and the improvement of computational thinking is closely associated with the process of solving problems by employing information technology tools. Though it starts with the demonstration of teachers, it is necessary to be adapted to students' thinking. Therefore, when the learning activities of units are designed, core concepts shall be focused on and the critical questions shall be raised for them to explore the core concepts and content. In doing so, students can be guided to gain an in-depth understanding of the core concepts through a continuous process.

RECOMMENDATION FOR COMPUTATIONAL THINKING EDUCATION

The prevalence of digital tools allows the program-driven technical tools to penetrate all aspects of people's lives, and the means of computing they apply internally are imperceptibly integrated into the problem-solving process by taking advantage of information technology. Therefore, not only could understanding the significant characteristics of digital tools and cultivating computational thinking achieve the effective use of information technology but it is also conducive to averting the danger of "being manipulated by technical tools."

As young people grow up in a digital environment, they develop proficiency in applying digital technologies and employing technical tools to better adapt to the digital environment. Nevertheless, in an information society, educators need to be aware that the widespread application of digital technology not only creates a diverse environment of technical application but also leads to the emergence of new core concepts, methods of solving problems with technology, and unique technical application standards. Therefore, as qualified "digital citizens," educators are supposed to gain understanding as to the core concepts, disciplinary methods, and corresponding communication forms of digital technology in practice; solve realistic problems by applying scientific methods in a sensible way; perform information activities in a responsible way in line with the behavioral standards that apply to the information society; and cultivate the basic literacy of survival, innovation, and development in the information society.

How can we plan well when faced with the future? In an era that is characterized by artificial intelligence, it is necessary for us to place the focus on the development of new computational thinking skills. The pressure coming from business and industry to transform has prompted all countries to shift their attention to the transformation that AI technologies can achieve in different fields. Artificial intelligence has been made a new focal point of international competition. In the meantime, there is a new opportunity for development presented to all people around the world (McArthur, Lewis, and Bishary 2005). As a major developing country, China is viewing its resources from a different perspective—whether human, natural, data, or intelligence—and the relationships between them (Jiang et al. 2017). Elementary education plays a crucial role in solving the question of what kind of people society wants to create, and AI has a substantial impact. Training people who are irreplaceable by AI is a critical problem that needs addressing. By learning an AI course module, students can understand the concepts and historical development of artificial intelligence while developing the capability to describe the process of developing common AI algorithms. Through a module where they develop their own simple intelligent technology applications, they can acquire hands-on experience with techniques and learn about the basic process to design and create a simple intelligent system, which is conducive to reinforcing their sense of responsibility in applying intelligent technologies to serve humans. Through STEM education, the combination of AI and the exploration of other courses can allow students to develop the capability of interdisciplinary thinking, while the basic mode of thinking behind AI can exert influence on students for their academic study and daily life.

However, the content of AI involves a wide range of subjects and is still in the process of exploration and improvement. This affects the development of curricula for primary and middle school students and may also lead to the lack of cohesion of artificial intelligence courses and the disconnection of course contents from the actual educational goals. To develop artificial intelligence education, it is necessary to make clear the practicality and systems of the educational setting. Artificial intelligence education in primary and secondary schools should focus on cultivating students' information literacy, aiming at improving students' understanding of information awareness, computational thinking, digital learning, and innovation. Attention should be paid to the laws of cognitive development, and the value of science and technology should be highlighted. In China, in recent years, relevant policies have been introduced focusing on artificial intelligence education in primary and secondary schools, and a series of artificial intelligence textbooks covering primary schools as well as junior and senior high schools have been published. The textbooks attach importance to the consistency of the content while setting different teaching objectives and strategies according to different sections.

What is the most urgent implementation path for computational thinking education for all? Firstly, "digital indigenous" does not automatically become a digital citizen. It is wrong to assume that the next generation can automatically become a qualified digital citizen by living in an information-rich environment, with thinking and ethics included (Ginsburg 2008). There are various aspects of information literacy that need to be taught in a systemic way. Secondly, computational thinking is regarded as an essential part of literacy for citizens involved in any work in the future. At present, the most feasible way to implement the curriculum is through programming and algorithmic training assisted by different tools for the cultivation of computational thinking. The programming training emphasized by the new curriculum standard is not limited to simple programming—not programming for the sake of programming—but developing students' computational thinking through programming. Therefore, it is just the starting point for the textbook reform, teaching reform, teacher education reform, and evaluation reform that we will need in the future. Nowadays, in the transformation of information technology that is improving the living standards for Chinese people, the reform to education and teaching has been deepened on a continued basis so that every single student is entitled to receive education in information technology.

For middle schools and primary schools, information technology courses provide the basic way to deliver information technology education. They need to be adaptive to the characteristics of the times so as to accomplish the crucial task of developing computational thinking for students and cultivating qualified digital natives.

REFERENCES

Brennan, Karen, and Mitchel Resnick. 2012. "New Frameworks for Studying and Assessing the Development of Computational Thinking." In *Proceedings of the 2012*

Annual Meeting of the American Educational Research Association, edited by American Educational Research Association, 1–25. https://web.media.mit.edu/~kbrennan/files/Brennan_Resnick_AERA2012_CT.pdf..

Brown, Neil C., Sue Sentance, Tom Crick, and Simon Humphreys. 2014. "Restart: The Resurgence of Computer Science in UK Schools." *ACM Transactions on Computing Education (TOCE)* 14, no. 2: 9.

Chai, Ching Sing, Joyce Hwee Ling Koh, and Chin-Chung Tsai. 2013. "A Review of Technological Pedagogical Content Knowledge." *Journal of Educational Technology & Society* 16, no. 2: 31–51.

Falkner, Katrina, Rebecca Vivian, and Nickolas Falkner. 2014. "The Australian Digital Technologies Curriculum: Challenge and Opportunity." In *Proceedings of the Sixteenth Australasian Computing Education Conference*, edited by Jacqueline Whalley and Daryl D'Souza, 148. Australia: Australian Computer Society.

Ginsburg, Faye. 2008. "Rethinking the Digital Age." In *The Media and Social Theory*, edited by David Hesmondhalgh and Jason Toynbee, 141–158. London: Routledge.

Grover, Shuchi, and Roy Pea. 2013. "Computational Thinking in K–12: A Review of the State of the Field." *Educational Researcher* 42, no. 1: 38–43.

Heintz, Fredrik, Linda Mannila, and Tommy Färnqvist. 2016. "A Review of Models for Introducing Computational Thinking, Computer Science and Computing in K-12 Education." In *Proceedings of 2016 IEEE Frontiers in Education Conference (FIE)*, edited by Steve Frezza and Dipo Onipede. Piscataway, NJ: Institute of Electrical and Electronics Engineers.

Jiang, Fei, Yong Jiang, Hui Zhi, Yi Dong, Hao Li, Sufeng Ma, Yilong Wang, Qiang Dong, Haipeng Shen, and Yongjun Wang. 2017. "Artificial Intelligence in Healthcare: Past, Present and Future." *Stroke and Vascular Neurology* 2, no. 4: 230–243.

Love, T. S., and Greg J. Strimel. 2016. "Computer Science and Technology and Engineering Education: A Content Analysis of Standards and Curricular Resources." *Journal of Technology Studies* 42, no. 2, 76–89.

Lye, Sze Yee, and Joyce Hwee Ling Koh. 2014. "Review on Teaching and Learning of Computational Thinking through Programming: What Is Next for K-12?" *Computers in Human Behavior* 41, 51–61.

Mannila, Linda, Valentina Dagiene, Barbara Demo, Natasa Grgurina, Claudio Mirolo, Lennart Rolandsson, and Amber Settle. 2014. "Computational Thinking in K-9 Education." In *Proceedings of the Working Group Reports of the 2014 on Innovation & Technology in Computer Science Education Conference.*, edited by Alison Clear and Raymond Lister. New York: Association for Computing Machinery.

McArthur, David, Matthew Lewis, and Miriam Bishary. 2005. "The Roles of Artificial Intelligence in Education: Current Progress and Future Prospects." *Journal of Educational Technology* 1, no. 4: 42–80.

Nouri, Jalal, Lechen Zhang, Linda Mannila, and Eva Norén. 2019. "Development of Computational Thinking, Digital Competence and 21st Century Skills When Learning Programming in K-9." *Education Inquiry* 11, no. 1: 1–17.

Papert, Seymour. 1990. "A Critique of Technocentrism in Thinking about the School of the Future." *M.I.T. Media Lab Epistemology and Learning Memo*, no. 2 (September).

Ravenscroft, Andrew. 2001. "Designing E-learning Interactions in the 21st Century: Revisiting and Rethinking the Role of Theory." *European Journal of Education* 36, no. 2: 133–156.

Wang, Tao. 2019. "Competence for Students' Future: Curriculum Change and Policy Redesign in China." *ECNU Review of Education* 2, no. 2, 234–245.

Wing, Jeannette M. 2008. "Computational Thinking and Thinking about Computing." *Philosophical Transactions of the Royal Society A: Mathematical, Physical and Engineering Sciences* 366, no.1881: 3717–3725.

4

COMPUTATIONAL THINKING AND THE NEW CURRICULUM STANDARDS OF INFORMATION TECHNOLOGY FOR SENIOR HIGH SCHOOLS IN CHINA

Ronghuai Huang, Junfeng Yang, Guangde Xiao, and Hui Zhang

CHALLENGE FOR IT EDUCATION

As information and communication technology (ICT) spreads rapidly around the world, digital technologies have influenced the way in which we learn, live, and work. Furthermore, in recent decades, intelligent technologies, including artificial intelligence (AI) and emerging advanced technologies empowered with AI techniques, such as the Internet of Things (IoT), virtual reality (VR)/augmented reality (AR) enhanced by AI and intelligent terminal devices, have risen to the forefront of public discussion. The impact of digital technologies is already being felt across all sectors of society, such as transportation (e.g., smart cars), health care, the education of low-resource communities, public safety and security, employment and the workplace, and entertainment (Stone et al. 2016). Digital technologies not only enable societal changes but people also depend on technologies to cope with the information society. The new generation of students is often labeled "digital natives" or the "Net generation." However, these students' ICT use is much more limited in scope than originally portrayed both in the Western and Eastern context (Aesaert and van Braak 2015; Li and Ranieri 2010). Consequently, it is essential for the education system to provide students with adequate competencies to cope with such a society. In the context of the digital age, several terms have been proposed over the last few decades. Digital competence, ICT literacy, and digital viability have all been used to identify and describe what students should be able to achieve within a digital context. School-based initiatives, in particular the information technology (IT) curriculum, is considered as an essential way to prepare students for the digital age.

Liu, Ren, Li, and Zhao (20018) emphasize that the IT curriculum in China has passed through three stages over the last forty years. The first stage (up to the 1980s), often called pilot computer education, was integrated into nineteen pilot high schools as an elective course in 1982, aiming to promote students' computer literacy, especially focusing on "the ability to use a spreadsheet and a word processor and to search the World Wide Web for information" (National Research Council 1999, 11). The second stage (1980s through 1990s), often called computer education, still focused on computer literacy and began to be integrated into IT application in primary and secondary schools. In the third and current stage (1990s through 2000s), the Ministry of Education released the guidelines for ICT curriculum in primary and secondary schools (trial), and computer education was renamed "IT Education" (Liu, Ren, Li, and Zhao 2018). By 2015, 100 percent of high schools, 95 percent of junior high schools, and 50 percent of primary schools had already set IT as a compulsory subject.

However, the K–12 IT curriculum in China currently faces many challenges. Firstly, the education objective still focuses on fostering students' basic technical skills (Liu, Ren, Li, and Zhao 2018). In the context of the digital age, the objective of the IT curriculum should not only cultivate students' ability in applying technology but also include problem-solving ability and understanding the principle of technology. Secondly, the framework and content of IT curriculum standards need to be revised according to current technology development (Dong, Liu, and Qian 2014). Technologies, especially intelligent technologies, are taking over repetitive work and even changing the very nature of employment. It is essential that the younger generation has the opportunity to develop an accurate understanding of technology—what it is, how it works, and how it might impact their lives—to equip them with the necessary "digital survival skills" for the workplaces of the future. Therefore, the content of the IT curriculum should be revised according to technology development. Thirdly, the teaching approach deployed to develop students' ICT literacy mainly focuses on introducing technical knowledge and exercising technical skills (Xiao and Huang 2016). More effective methods should be implemented to foster students' ICT literacy, such as inquiry-based learning, problem-based learning, and collaborative learning.

NEW CURRICULUM STANDARDS OF IT FOR SENIOR HIGH SCHOOLS IN CHINA

To cope with the previously mentioned challenges and cultivate digital citizens in China, the Ministry of Education in China has released the new curriculum standards of IT for senior high school (hereafter referred to as New Standards) in 2017.

THE FOUR-KEY LITERACY OF THE NEW STANDARDS

The aim of the New Standards is to cultivate people with ICT literacy who have information consciousness, computational thinking skills, digital learning and innovative ability, and responsibility for the information society, which is called the four-key literacy of the standards (see figure 4.1).

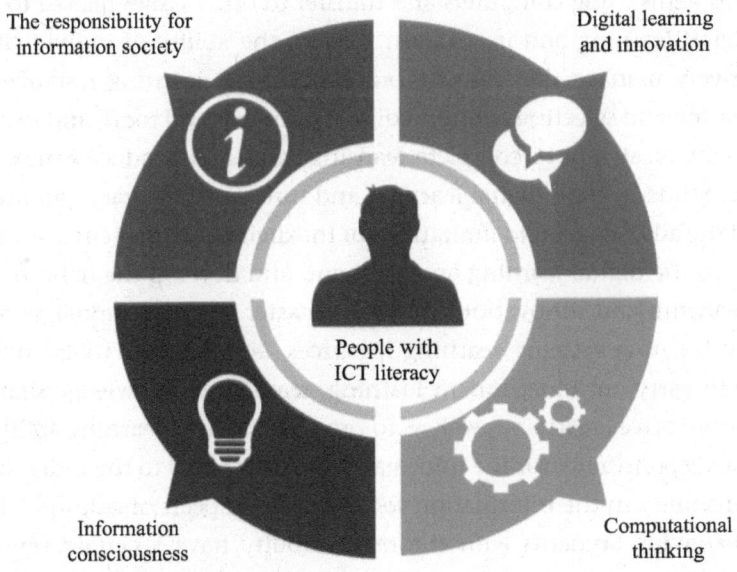

The responsibility for information society

Digital learning and innovation

People with ICT literacy

Information consciousness

Computational thinking

4.1 The four-key literacy.

Information consciousness refers to the individual's sensitivity to information and his/her judgment of the value of information. Students who have information consciousness can consciously and actively seek appropriate ways to obtain and process information according to the need for solving the problem, and they will be able to notice the changes in the information to analyze the information contained in the data. Students with information consciousness could also use effective strategies to make reasonable judgments on the reliability of the source of information, the accuracy of the content, and the possible impact of the information to provide a reference for solving the problem. Students with information consciousness would be more likely to share information with team members to solve problems in a collaborative way.

Computational thinking refers to a series of thinking activities in the process of finding solutions by using the methods of computer science. Students with computational thinking can define problems, abstract features, establish structural models, and organize data reasonably in the way that computers can handle; they can form solutions by using reasonable algorithms through judging, analyzing, and synthesizing various information resources and summarize the process and methods of solving problems using computers and transfer to other issues related to it.

Digital learning and innovation refer to the ability of individuals to effectively manage the learning process and the learning resources by evaluating and selecting common digital resources and tools, and to solve problems creatively to complete learning tasks and produce innovative works. Students with digital learning and innovation literacy can understand the advantages and limitations of the digital learning environment, adapt to the digital learning environment, and develop the habit of digital learning and innovation; they can master the operational skills of digital learning systems, learning resources, and learning tools and use these to carry out independent learning, teamwork, knowledge sharing, and innovative creation as well as to promote lifelong learning ability.

The responsibility for the information society refers to the individual's responsibility in the information society in the aspects of culture, ethics, and behavior. Students with the responsibility have a certain sense of information security, and they will abide by information laws and regulations, abide by the moral and ethical norms of the information society, and abide by public norms in real and virtual spaces. They can effectively safeguard the legitimate rights and interests of themselves in information activities, and they can actively safeguard the legitimate rights and interests of others. They will pay attention to the environmental and human problems brought about by the information technology revolution, hold positive learning attitudes to new ideas and new things arising from IT innovation, make judgments rationally, and act responsibly.

Each part of the four-key literacy has three levels, and the rules and conditions of each level are defined in the standard, which could be used to determine the level achieved by high school students after completing the IT course. To solve the problem of poor connectivity between high school and primary or middle school, the New Standards specifically determines the preparatory level of each key literacy, which points out the level of four-key literacy that students should have when they enter high school. The identification of the preparatory level provides a reference standard for the IT course in K–9, which makes a clear distinction between the task of the IT course in primary and junior school and in high school.

Taking computational thinking as an example, the rules and conditions of the preparatory level and other three levels are shown in table 4.1.

THE CONTENT OF THE IT CURRICULUM IN THE STANDARDS

To select suitable content for the high school information technology curriculum, the new curriculum standard takes the key literacy as a guide and defines the general concept of information technology discipline. The new curriculum identifies four core concepts, including data, computing, information systems, and the information society. Data is the carrier of information, which is the basic processing and calculation object for information technology, and the effective processing of various types of data produces development information technology. The algorithm is the basis of the data processing method, affecting the effectiveness of data processing, which is also the basis of developing new information technology. The information system is the materialization of information technology that comprises the basic objects for people to use, recognize, and understand information technology. The information

Table 4.1   The rules and conditions of the levels for computational thinking

Level	Computational thinking
Preparatory level	• understand the advantages of digitalized information in daily life • be able to identify key features for a given simple task and draw a flowchart for completing the task • understand the value, processes, and tools of processing information and be able to select the right tools according to the needs
Level 1	• be able to conduct a needs analysis for a given task and identify the key issues that need to be addressed • be able to extract the basic features of the problem, abstract the problem, and express the problem in a formalized way • be able to use basic algorithms to design solutions to problems and use a programming language or other digital tools to implement this solution • be able to use the appropriate digital tools or methods to obtain, organize, analyze data, and transfer the process to other related issues
Level 2	• be able to use formalized methods to describe problems and use modular and systematic methods to design solutions to complex problems • be able to correctly distinguish the various data involved in the problem-solving process and use the appropriate data type for representation • be able to design or select the appropriate algorithm for different modules and use the programming language or other digital tools to implement the functions of each module • be able to integrate the functionality of each module with the appropriate development platform to achieve a total solution
Level 3	• be able to use the principles of information system design to conduct a more comprehensive evaluation for information-based solutions and use appropriate methods to iteratively optimize solutions • be able to transfer the process of using IT to solve problems to the solution of other related problems in daily life and the learning process

society is the result of the influence of information technology on human society. Recognizing and understanding the social form of the economic, political, and cultural activities of the information society, adapting to and obeying the corresponding rules, is the premise of the survival and effective self-development in the information society.

According to the key literacy and big concept of information technology discipline, the New Standards determines the structure of the curriculum content, as shown in table 4.2.

The required part of the information technology course aims to realize the all-around development of students, and the selective required part aims to promote the deep development of students' personalities. The foundation content emphasizes that students can understand the essence of information technology from the theoretical level and scientific point of view, which is the common basis of all students' learning, and the contemporary content focuses on making students understand the latest technology and mastering the basic methods of application, design, and development of new technologies to form the consciousness of innovative application of new technologies.

Both practice and theory are emphasized in the new IT curriculum. In the practical aspect, the emphasis is placed on developing students' practical skills to adapt to the needs of the new era, such as 3D design and creativity, open-source hardware project design (engineering), and mobile application design (network). On the theoretical aspect, the content

Table 4.2 Content structure of the IT curriculum

Big concepts	Required part	Selective required part	Selective part	Character
Data algorithm	Data and computing	• Data and data structure • Data management and analysis • Basics of artificial intelligence	Basics of algorithm	Foundation
Information system Information society	Information system and society	• Basics of network • 3D design and innovation • Open-source hardware design	Design for mobile app	Contemporary

of data and computing, information systems and society, data structure, data management and analysis, artificial intelligence preliminary, and algorithmic preliminary are set to cultivate students' computational thinking ability.

COMPUTATIONAL THINKING AND KEY LITERACY OF IT

The concept of computational thinking (CT) was originally introduced in Seymour Papert's book *Mindstorms: Children, Computers, and Powerful Ideas* (Papert 1980) with his creation of the programming language LOGO. Papert (1980) refers to CT primarily as the relationship between programming and thinking skills. Unlike Papert's, many definitions of CT in the twenty-first century emphasize the concepts that are commonly engaged in programming or computer science (Zhang and Nouri 2019). In one of the most popular definitions, Jeannette M. Wing (2006) defines computational thinking as "solving problems, designing systems, and understanding human behavior, by drawing on the concepts fundamental to computer science, which includes a range of mental tools that reflect the breadth of the field of computer science." Wing argues that computational thinking is a fundamental skill for everyone, not just for computer scientists, and is a basic literacy of people's understanding of problems, analyzing problems, and proposing solutions in the information society. Computational thinking is essentially a kind of thinking method or thinking activity that can solve problems flexibly by using computational tools and methods. Its value is not only reflected in the effective overcoming of the knowledge gap, building an interdisciplinary dialogue bridge, but more importantly, it plays an irreplaceable role in promoting the overall development and lifelong development of human beings (Fan, Zhang, and Li 2018).

From the development trend of the international information technology curriculum, the content of scientific and principled content in information technology courses has been greatly increased. Computational thinking has been incorporated into the Computer Science Teachers Association (CSTA) K–12 standard (2011 revision), the UK's new curriculum plan for 2013 is to develop computer science as an important teaching element of information technology teaching, and the new curriculum program developed in Australia in 2015 has made computational thinking an important part. In addition, the European Union core literacy, the United States's "21st century skills," "Horizon Report," and other international research results as well as the Programme for International Student Assessment (PISA); science, technology, engineering, arts, and mathematics (STEAM) education; and the concept and practice of the content of the creator education have greatly emphasized the theoretical and scientific content of the information technology curriculum.

Computational thinking has become an important factor for people to adapt to the survival and development of the information society in the future as well as to build their own development. Cultivating students' computational thinking is the current requirement and task in developing the education system. After considering the problems in the implementation of China's information technology curriculum and the development trend of international information technology education, the high school information technology curriculum will train students' computational thinking as an important goal of the curriculum and greatly increase the content of computer science to train students in preliminary computational thinking.

The curriculum of information technology in senior high school should be aimed to cultivate students' abilities at a high level. High school is still in the basic education stage and connects with higher education, so the information technology curriculum in high school needs to train students in the most critical and necessary basic literacy, rather than the professional ability at the university level. Based on the consideration of various conditions, computational thinking is identified as one of the core literacies out of the four-key literacy in the New Standards.

THE BASIC METHODS OF TEACHING CT

In the information technology classroom, when cultivating students' computational thinking abilities, the instruction should be guided by problem-solving, and the teacher should guide students to think and analyze problems based on the thinking methods in computer science (Huang and Xiao 2019). In the instructional practice, teachers are required to master the following three main points.

First, the instruction should begin with setting the right context for problems. At the beginning of a class, an authentic (real-world) problem should be given to students to solve, with the background and conditions of the problem clearly described. The reason the authentic problem should be given to students first is that computational thinking is problem-oriented and students need to concentrate on thinking about the real-world problems with the thinking methods of computer science. Unlike the computer professional education that imparts professional knowledge, students need to master the thinking methods instead of computer expertise. It should be noted that the problem should not necessarily be difficult and complicated, but there should be a clear computer science–oriented feature that can provide a solution from a mathematical and engineering perspective rather than setting a simple application of a particular tool software as a solution.

Second, the teacher should guide students to explore solutions to the problem by using the method of computer science. In the New Curriculum Standard, the general process of problem-solving with computational thinking is clearly stated. The first step is to use ideological methods in computer science to define problems, abstract features, build structural models, and organize data; the second step is to form problem-solving schemes by using reasonable algorithms through judging, analyzing, and synthesizing all kinds of information resources; the third step is to summarize the processes and methods of using computers to solve problems and apply them to other problem-solving related to them. In this process, students should use the thinking methods of computer science to characterize the problem, abstract features, and build models as well as to use relevant resources and algorithms to produce solutions, and each step is inseparable from computer science. However, students do not naturally understand computer science, and therefore instructors should teach the computer science knowledge and thinking methods involved in the problem-solving process to help students lay the necessary foundation. Teachers' guidance is the necessary condition for students to explore solutions smoothly, and teachers need to analyze and design the process of problem-solving in advance, closely track students' problems and progress in each process, and give the necessary guidance and help in a timely manner. The guidance should enable students to gain a deep understanding of the internal mechanisms of the solution to promote knowledge transfer.

Third, the relationship between computational thinking and programming should be correctly handled. In the process of finding a solution to a problem, writing the necessary programs and turning them into an automated solution is a necessary part of problem-solving based on computational thinking, which is likely to take up a lot of teaching time. If the teaching methods are not appropriate, or the emphasis is inappropriate, it is likely that the teaching of computational thinking will be transformed into programming-oriented teaching. The key is the orientation of teaching: the teaching of computational thinking is problem-oriented with the goal of finally obtaining the solution to the problem, while the goal of programming is usually to master a programming language, which is oriented by learning grammar knowledge.

The New Standards advocates project-based learning and comprehensively promotes the formation of the four-key literacy in the process of students' participating in the project. Project-based learning (PBL) is a teaching method in which students learn by actively engaging in real-world and personally meaningful projects. PBL usually requires authentic learning contents that come from real life. Under the guidance of teachers, students engage in solving real-world problems and in experiencing the complexity, comprehensiveness, and integrity of problem-solving.

PBL activities often require students to use certain technical tools and research methods to solve the problems. There is a natural advantage to using the PBL method in the IT course. First of all, the problems in the IT course often come from real life, and many of them are closely related to students' daily lives and learning. Second, the IT courses are often taught in technology-rich environments that can provide proper technical tools and learning environments. Third, the process of completing the project is also the process of applying, improving, and innovating IT, and it is also the process of improving computational thinking.

THE ENVIRONMENTS FOR TEACHING CT

According to the traditional understanding, many people think that the teaching environment for IT courses is a multimedia network classroom, which was previously sufficient for implementing IT courses. However, the multimedia network classrooms cannot meet the requirements of the new IT curriculum standards. In the IT course, students should go through a series process of thinking, identifying problems, proposing solutions, exploring solutions, evaluating solutions, and migrating applications, and they gradually develop the four-key literacy in the process of solving problems. The environment should support the student's learning process and also support the teacher's project-based instruction, with the aim to cultivate the four-key literacy.

As shown in figure 4.2, the learning environment to support PBL in cultivating computational thinking should include new textbooks, an online community for teachers, open sources and toolkits, an innovation laboratory, a cyberlearning space, and open educational resources (OER). New textbooks considering the New Standards should be adopted to guide students to experience the process of computational thinking through solving problems and accomplishing projects. An online community for IT teachers is important to provide a platform for communication and reflection on the teaching methods and solving problems. Open software and toolkits are important tools for conducting learning activities for computational thinking, with which students and teachers can work together to solve real-world problems.

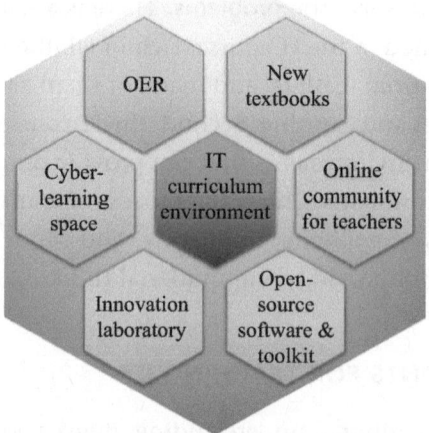

4.2 The IT curriculum environment to support PBL.

The innovative laboratory provides a physical learning environment where students can explore problems, inquire about solutions, and make artefacts by using different kinds of software and tools. Information technology can also be used to create a cyberlearning space for students with convenient applications, abundant resources, reliable contents, and environmental security. The cyberlearning space combined with physical space can provide students an environment suitable for self-directed and inquiry learning, provide convenient ways for teacher-student communication, and also enhance the teaching methods. In the technology-enhanced learning environments, students can experience the digitalized learning process, which promotes the habit of lifelong learning. The Internet should be best utilized to promote the sustainable development OER, with the mechanism of transforming students' learning processes to learning resources, guiding students to become both the user and builder of learning resources.

ACKNOWLEDGMENTS

This research work is supported by Zhejiang Provincial Philosophy and Social Planning Project: Design Learning Spaces for Digital Generation Students (19ZJQN21YB).

REFERENCES

Aesaert, Koen, and Johan van Braak. 2015. "Gender and Socioeconomic Related Differences in Performance Based ICT Competences." *Computers & Education* 84: 8–25. https://doi.org/10.1016/j.compedu.2014.12.017.

Dong, Y., X. Liu, and S. Qian. 2014. "The Last Development Trends and Implication of International ICT Course in Primary and Secondary School." *China Educational Technology* 2: 23–26.

Fan, W., Y. Zhang, and Y. Li. 2018. "Review of Research and Development of Computational Thinking Abroad." *Journal of Distance Education* 36, no. 2: 3–17.

Huang, R., and G. Xiao. 2019. "Key Issues for Teaching in Implementing New Curriculum Standards of Information Technology Courses." *Digitalized Teaching in Primary and Middle School* 2: 5–8.

Li, Yan, and Maria Ranieri. 2010. "Are 'Digital Natives' Really Digitally Competent? A Study on Chinese Teenagers." *British Journal of Educational Technology* 41, no. 6: 1029–1042.

Liu, Ruixue, Youqen Ren, Feng Li, and Jian Zhao. 2018. "Entering a New Era: History Achievements, Challenges and Reform Strategies of Information Technology Education in Primary and Secondary Schools of China." *Modern Educational Technology* 28, no. 6: 17–24.

National Research Council (NRC). 1999. *Being Fluent with Information Technology*. Washington, DC: National Academy Press.

Papert, Seymour. 1980. *Mindstorms: Children, Computers, and Powerful Ideas*. New York: Basic Books.

Stone, Peter, Rodney Brooks, Erik Brynjolfsson, Ryan Calo, Oren Etzioni, Greg Hager, Julia Hirschberg, et al. 2016. *"Artificial Intelligence and Life in 2030." One Hundred Year Study on Artificial Intelligence: Report of the 2015–2016 Study Panel*. Stanford, CA: Stanford University. http://ai100.stanford.edu/2016-report.

Wing, Jeannette M. 2006. "Computational Thinking." *Communications of the ACM* 49, no. 3: 33.

Xiao, G., and R. Huang. 2016. "Problems in Implementing Information Technology Curriculum of High School and the Strategies of New Curriculum Standards." *China Educational Technology* 12: 10–5.

Zhang, LeChen, and Jalal Nouri. 2019. "A Systematic Review of Learning by Computational Thinking through Scratch in K-9." *Computers & Education* 141: 103607. https://doi.org/10.1016/j.compedu.2019.103607.

5

COMPUTATIONAL THINKING CURRICULA IN AUSTRALIA AND NEW ZEALAND

Tim Bell, Rebecca Vivian, and Katrina Falkner

INTRODUCTION

Australia and New Zealand (NZ) are neighboring countries that have independent education systems, but both have introduced the subject of "digital technologies" over a similar period, and the process has been in political and cultural contexts that have a lot in common. Both countries have the subject situated within a curriculum area that covers technology in general, which has a focus on design, innovation, and end-users.

This chapter explores similarities and differences through the lens of categories based on the Darmstadt Model, a framework that provides a category system for comparing and contrasting computer science (CS) education across regional or national boundaries (Hubwieser et al. 2011). We have selected aspects that are particularly relevant to understanding how the curricula have evolved in these two countries.

EDUCATIONAL SYSTEM

New Zealand and Australia have independent education systems, and in fact within Australia states and territories can determine how education is run. Consequently, the countries (and states/territories) are not exactly the same in their approach to education although there are considerable commonalities. In both countries, the new learning area in the curricu-lum that correlates to computing is called "digital technologies" (DT). In these curricula, DT includes substantial elements relating to both com-puter science (or computational thinking) and using digital devices as a tool. An extensive comparison of the curricula of the two countries has been published by ACARA (2019).

NEW ZEALAND

The New Zealand pre-tertiary school system generally has thirteen years of schooling, referred to as "year 1" to "year 13." Children typically start at year 1 in primary school on their fifth birthday. High school usually starts at year 9, and from years 11 to 13 students sit for school-leaving assessments, with the "National Certificate of Educational Achievement" (NCEA) being the most common national qualification that covers those three years (New Zealand Qualifications Authority 2021).

New Zealand is a bicultural country, and the New Zealand school sys-tem operates under two main curricula: the "New Zealand Curriculum" (NZC) for English medium instruction and Te Marautanga o Aotearoa (TMoA) for Māori medium instruction (New Zealand Ministry of Educa-tion 2021a). These curricula are not simply translations of each other but organize knowledge differently with different emphasis. Nevertheless, computing falls under the area of "technology" in the NZC, and under "hangarau" (which can be translated as "technology") in the TMoA. Within these areas, they are slightly different but have a lot in common; a key difference is the context in which they are taught. In this chapter, we will focus on the English-medium New Zealand curriculum.

The "front end" of the NZC includes values and key competencies that are expected to be developed and promoted through schools' delivery of the curriculum. The key competencies are thinking; relating to others; using language, symbols, and texts; managing self; and participating and contributing. There are clear opportunities to practice these in many aspects of computing.

The subject content of the NZC is articulated through eight "learning areas." One of these learning areas is "technology" (New Zealand Ministry of Education 2021b), the other seven being typical curriculum areas such as "mathematics and statistics," "the arts," and "science." Within the technology learning area, two out of its five "technological areas" (figure 5.1) are grouped as "digital technologies" (DT); these are "computational thinking for digital technologies" (CTDT) and "designing and developing digital outcomes" (DDDO).

Each of CTDT and DDDO has a series of "progress outcomes" (New Zealand Ministry of Education 2021b) that show the stages a student should go through as they progress in these areas. The progress outcomes essentially define the curriculum, and so they indicate the level of detail that is provided to schools on what they should teach and assess. These are described in more detail in the "Intentions" section later in this chapter.

Supporting documentation including exemplars is provided, although the New Zealand education system has a strong emphasis on "local

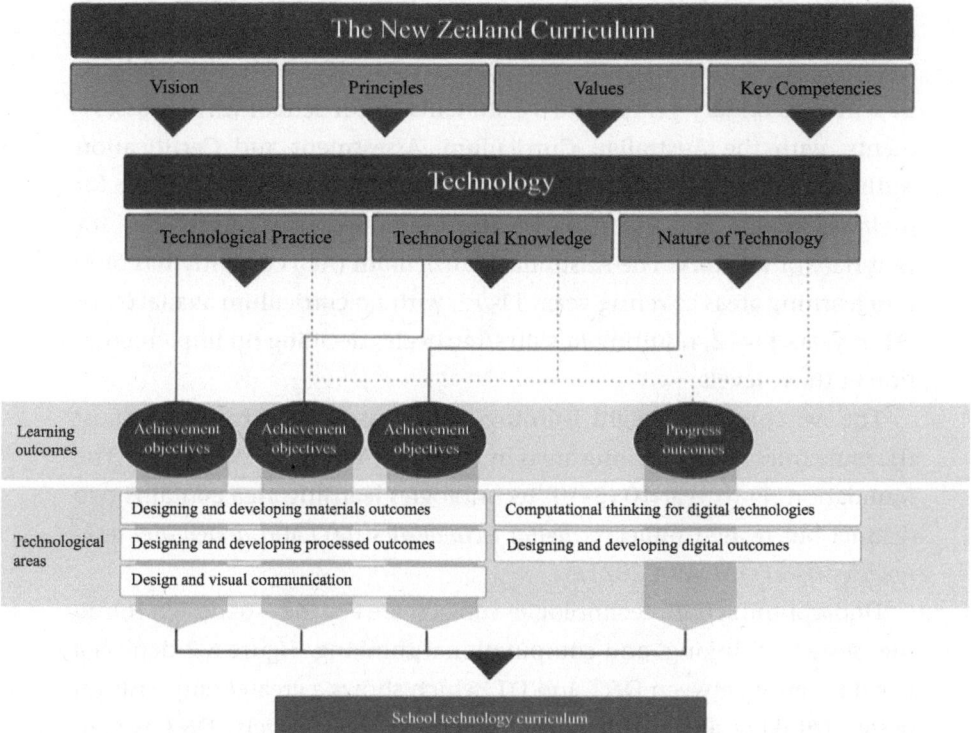

5.1 Structure of the technology learning area in the New Zealand curriculum (from nzcurriculum.tki.org.nz) (New Zealand Ministry of Education 2021i).

curriculum" (New Zealand Ministry of Education 2021j), where schools work out how to deliver the material in the most appropriate way for their community. Thus, the NZC is not highly specified like some curricula, and a lot of autonomy is given to schools and teachers to choose when it is delivered, what textbooks or resources are used, and how it is integrated with other learning areas.

AUSTRALIA

Similar to New Zealand, the Australian pre-tertiary school system also typically has thirteen years of schooling, with a preliminary year, referred to in this chapter as "foundation," to align to the Australian Curriculum (states vary in their naming of the initial year), followed by "year 1" through "year 12" (ACARA 2021a). Children typically begin their first year of primary school at five or six years of age. Secondary school (or high school) usually starts in year 7 or 8 (depending on the state). While a curriculum has been developed up until year 10, years 9 to 10 are elective subjects. In senior secondary years 11 to 12 students sit for school-leaving assessments, with the Australian Curriculum, Assessment and Certification Authority (ACACA agency) in each state or territory being responsible for their assessment and reporting for the Senior Secondary Certificate for that state or territory. The Australian Curriculum (AC) currently has only four learning areas covering years 11–12, with no curriculum available for DT in years 11–12, resulting in states/territories deciding on implementation at those levels.

The AC consists of eight learning areas that happen to have almost the same titles as the learning areas in the NZC (ACARA 2019, table 1). The foundation (F) to year 10 (F–10) technologies learning area contains two distinct but related subjects: *digital technologies (DT)* and *design and technologies (D&T)* (ACARA 2021a).

Underpinning both technologies subjects are key ideas of design thinking, systems thinking, and computational thinking. Figure 5.2 depicts a key difference between D&T and DT, which shows a greater emphasis on design thinking and computational thinking, respectively. D&T is concerned with contexts related to engineering, food and fibre, food specializations, and materials and technologies specializations. DT is the subject area that includes computing content and is the focus of this chapter.

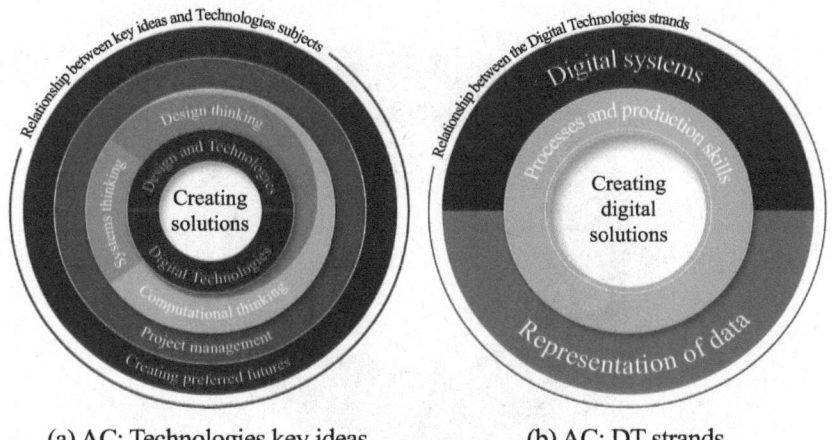

(a) AC: Technologies key ideas   (b) AC: DT strands

5.2   AC: DT key ideas and strands (from ACARA 2021c).

The AC: DT subject includes two related strands (figure 5.2b): "knowledge and understanding," which is "the information system components of data, and digital systems (hardware, software and networks)," and "processes and production skills," which involve "using digital systems to create ideas and information, and to define, design and implement digital solutions, and evaluate these solutions and existing information systems against specified criteria."

The Australian government is committed to "closing the gap" and empowering Aboriginal and Torres Strait Islander students to reach their potential (Education Services Australia 2019). Unlike New Zealands's Te Marautanga o Aotearoa for Māori medium instruction, Australia does not have an explicit curriculum for Aboriginal and Torres Strait Islander education. Australia has taken the approach to include "Aboriginal and Torres Strait Islander Histories and Cultures" as one of the three identified cross-curriculum priorities that aims to highlight opportunities across the curriculum for all students to deepen their knowledge with a framework of First Nations Peoples, Culture and Country/Place (see figure 5.3). The cross-curriculum priorities are presented as guiding elaborations, embedded within learning areas alongside some content descriptors. As an example, in years 5–6 an elaboration suggests: "comparing past and present information systems in terms of economic, environmental and social sustainability, including those of Aboriginal and Torres Strait Islander Peoples."

5.3 Aboriginal and Torres Strait Islander Histories and Cultures cross-curriculum priority framework (ACARA 2021d).

The implementation of the F–10 AC is the responsibility of state and territory school and curriculum authorities (ACARA 2021e), who decide how and when the AC is implemented in their jurisdiction, including if they will develop their own version. All states or territories have either chosen to use this curriculum or a very similar version, except New South Wales (NSW) where they have developed their own syllabus based on the AC, featuring "digital technologies" within the years F–6 science and technology and years 7–10 information and software technology syllabuses (NSW Education Standards Authority 2021). The Australian Curriculum was being reviewed in 2020–2021, and version nine of the AC was published on a new website in early 2022.

SOCIOCULTURAL AND POLICY FACTORS

The development of digital technologies in both countries has come from a mixture of grassroots concerns from educators frustrated about the place of computing in the curriculum, and a push from industry to grow the diversity and number of qualified people to address the lack of talent available. As has been the case in most countries, this has been set against the typical issues of decision makers not understanding the difference between *using* and *creating* digital technologies, the difficulty of finding space in an already full curriculum for new topics, and the capacity to support teachers to deliver this new material.

NEW ZEALAND

The current NZC was developed in the 1990s, and at the time included the "technology" learning area. However, this had a broad focus—technology is defined in the curriculum as "intervention by design," and the curriculum noted that "technological areas include structural, control, food, and information and communications technology and biotechnology. Relevant contexts can be as varied as computer game software, food products, worm farming, security systems, costumes and stage props, signage, and taonga [prized objects]." The curriculum emphasized general principles rather than particular types of technology, and the key elements were (and remain) three "strands": technological practice, technological knowledge, and nature of technology.

In 2008, concerns were expressed (particularly from industry and universities) about the lack of visibility of computing as a discipline and the confusion around teaching *with* digital devices rather than teaching *about* digital devices.

This led to a review that initially resulted in new standards being introduced to the NCEA assessments, which are typically used by students in their last three years of high school. Prior to this change, the only computing assessments that these students had access to were more focused on *using* computers and were "unit standards." These weren't useful for qualifications such as university entrance, so they were not attractive to students who might be interested in studying computer science in the future. A number of new "achievement standards" were added to the NCEA from 2011, and many students used these to engage with programming and other computer science topics (Bell, Andreae, and Robins 2014).

In 2014, the New Zealand government announced a project called "A Nation of Curious Minds—He Whenua Hihiri i te Mahara," which included the goal of "more science and technology-competent learners, and more choosing science, technology, engineering and mathematics (STEM)-related career pathways." One result of this was to develop new curriculum content that "will be available for all students from Year 1 to Year 13." The timeline was to publish the content at the end of 2017 for use from 2018, and "in term 1 2020, it will be expected that schools will be teaching the new content" (Curious Minds 2021).

This work resulted in the technology learning area being revised. This appeared as an addendum to the curriculum (New Zealand Ministry of Education 2021d) in which five "technological areas" are defined: designing and developing material outcomes, designing and developing processed outcomes, design and visual communication, designing and developing digital outcomes, and computational thinking for digital technologies; the latter two are grouped as "digital technologies."

Although all schools were "expected" to offer the new content by 2020, its introduction was hindered by factors described next that made it challenging to engage teachers with the new content.

A more detailed history of these changes can be found in Bell, Andreae, and Lambert (2010); Thompson and Bell (2013); Thompson, Bell, Andreae, and Robins (2013); Bell (2014); Bell, Andreae, and Robins (2014); Fox-Turnbull (2019); and Falloon (2020).

AUSTRALIA

On the fourteenth of October 2014, the Australian government released the Industry Innovation and Competitiveness Agenda (Australian Government 2014) that aimed to strengthen Australia's competitiveness, with a major announcement being the introduction of the "coding across the curriculum program" to enhance computer programming skills across the curriculum. Until this point, the teaching of computing as a discipline was largely left to tertiary study, with school curricula focused on the use of digital tools for learning and teaching. The AC: DT subject has introduced computing as a specific discipline area, with the effective use of technologies being captured as one of the seven general capabilities across the curriculum within the information and communication technology (ICT) capability (ACARA 2021f). The AC highlights opportunities where ICT capabilities can be used across subjects/learning areas, including for DT. There have been challenges in supporting teachers to understand the distinction between ICT capability and DT, which required targeted messaging in professional learning and resources. Consultation for the ICT capability opened in late April–July 2021 and includes a review of all aspects of the curriculum, including the proposed renaming of the general capability to digital literacy and revision of the learning continuum.

The ACARA is an independent national authority responsible for advice and delivery of the national curriculum, assessment, and reporting for all Australian education ministers, including development of the AC. The development of the AC was inspired by the 2008 Melbourne Declaration of Educational Goals for Young Australians (Ministerial Council on Education, Employment, Training, and Youth Affairs 2008), in which all Australian governments committed to building equity and excellence and for all young Australians to become successful learners, confident and creative individuals, and active and informed citizens. The declaration also called for the need for students to develop ICT skills in response to evolving technologies in society. In 2019, the new Alice Springs (Mparntwe) Education Declaration was approved by ministers of education, highlighting the STEM learning areas that are a key national focus for school education in Australia (Education Services Australia 2019).

The initial "shape of the F–12 AC" was first approved by the Australian Council of Commonwealth and state and territory education ministers in 2009 (ACARA 2021g). The paper reflected the position of the 2008 Melbourne Declaration. The conceptualization and development of the AC: technologies commenced in 2010 and underwent a rigorous consultation and writing process, with the curriculum being endorsed and published in December 2015. Final implementation of the AC: DT was to the responsibility of states and territories, with the support of ACARA.

Following five years of implementation, in 2020, education ministers agreed that it was timely to review the F–10 Australian Curriculum (ACARA 2021h), including broad stakeholder consultation, feedback from state and territory jurisdictions, and research benchmark comparisons against selected international curricula of Singapore, Finland, British Columbia and New Zealand. The review was completed in 2021This brings a critical point for the future of the AC: DT and the direction Australia will take moving forward—whether it be to remain the same or adapt.

TEACHER QUALIFICATION

Teacher capability is a key bottleneck that needs to be addressed to effectively introduce digital technologies as a subject. In New Zealand and Australia, programming and related knowledge were not a part of many teachers' education since it was not in the school curriculum and relatively few teachers had computing degrees.

NEW ZEALAND

In New Zealand, teachers aren't required to have a specific qualification to teach digital technologies and may have come into the subject from a variety of pathways. The traditional "technology" subject in schools grew out of subjects such as woodwork and cooking, which were replaced by areas with names such as "materials technology" and "food technology." This also included "information and communication technology," but ICT teachers often came from a background in teaching typing, and later on office software, so they were more focused on using the computer as a tool.

When DT was introduced to the NCEA qualification in 2011, sources of upskilling for teachers included a series of "CS4HS" (computer science for high school) workshops run by universities. CS4HS was supported by Google, Inc. and grew out of an outreach program in the US (Blum and Cortina 2007) that seeded many opportunities for teacher professional development in Australasia (Thompson and Bell 2013). Grassroots support arose quickly in the form of a professional teachers' association, originally called the "New Zealand Association of Digital, Information, Technology Teachers" (NZACDITT), which later changed its name to "Digital Technologies Teachers Aotearoa" (DTTA). Through this group, teachers supported each other to work with the new curriculum, and later this turned out to be valuable because it provided a voice and support for teachers when the DT curriculum content was being extended for more junior classes.

After the full (year 1 to 13) curriculum was introduced, the government announced a raft of initiatives to help teachers and schools grow their capacity in DT, including "$24 million of new money toward additional professional learning and development for teachers" (New Zealand Government 2017). A large initiative that this funded was the "Kia Takatū ā-Matihiko" (te reo Māori, which roughly translates to "Be Ready for Digital Technologies") project (New Zealand Ministry of Education 2021e), a three-year government-funded multiorganization project to provide a large range of professional development for teachers, including a self-review tool, an online community, curation of teaching resources, online training, and in-person meetups as well as programs supported by other organizations such as "Code Club for Teachers" (Code Club 2021) and events run through The Museum of New Zealand—Te Papa Tongarewa. The Kia Takatū ā-Matihiko program was bicultural, providing teacher support and resources in both English and te reo Māori, for both the digital technologies and hangarau matihiko curricula content. The government also funded related initiatives, including professional development providers to work in schools, which the schools could apply for, and a national competition called "123Tech" (ITP NZ 2021) for students. These initiatives reached many teachers, although a final analysis is yet to be reported.

Two factors interfered with these 2018 to 2020 initiatives. In 2019 there was considerable industrial action from teachers and then from principals, and because the initiatives often did not fund time out of school for teachers, it was difficult to expect them to participate fully. Then, in 2020, the Covid-19 pandemic meant that teachers were working under heavy workloads to implement distance teaching for their students, which again reduced their availability to participate. One positive effect of the pandemic was that professional development programs also went online, and this meant they could reach teachers in distant communities who may not have participated otherwise.

Starting from an indigenous framework enabled an equitable, accessible, and bicultural approach as the basis of the initiatives described previously, but two programs that specifically addressed equity issues during the rollout of the new curriculum were "Raranga Matihiko | Weaving Digital Futures" (Museum of New Zealand—Te Papa Tongarewa 2021) and "Digital Ignition | Māpura Matihiko" (Digital Ignition 2021), which were designed to be accessible for schools so that distance and resourcing issues could be overcome to enable students to engage with the new curriculum content. Raranga Matihiko ended up running a home-learning TV show when schools were closed due to Covid-19.

AUSTRALIA

There are 149,462 primary and 138,832 secondary full-time or equivalent teachers across 9,503 schools in Australia (ACARA 2021i). As in New Zealand, Australian primary school teachers are generalists, required to teach across all learning areas without subject qualifications; as a consequence, pre-service primary teacher programs are expected to cover all learning areas broadly. High school teachers do not need specialist qualifications to teach a subject, but some may have chosen to major in technologies (design and technologies and/or digital technologies). Some schools may have "specialist" teachers dedicated to that subject area and responsible for teaching that subject in all classes.

Most early efforts to introduce computing into schools prior to the launch of the AC: DT was via outreach efforts from universities and industry keen to engage more young people in computing study and career pathways. University outreach has been primarily led by CS departments, and like New Zealand, some of these efforts have also been supported through the CS4HS funding program offered by Google Australia, now called Educator PD grants (Google for Education 2021). Since 2011, Google's program has supported 188 grants to sixty-six organizations delivering CS PD to primary, secondary, and pre-service educators across Australia and New Zealand.

Coinciding with the AC: DT release, on 7 December 2015, the Australian government announced its National Innovation and Science Agenda (NISA), committing $1.1 billion over four years to twenty-four measures across research, innovation, and entrepreneurship, including $84 million toward inspiring Australian students in digital literacy and STEM from 2016–2019 (Australian Government Department of Industry, Science, Energy and Resources 2015) across fifteen initiatives. On 11 December 2015, the Australian Education Council endorsed the National STEM School Education Strategy 2016–2026, further bolstering this initiative. School initiatives from 2016–2020 that focused on equipping school students to create and use DT included:

- The University of Adelaide (UoA) DT Massive Open Online Course (MOOC) program—Providing online professional learning for teachers on implementing the AC: DT with professional learning support via project officers and a free digital technology lending library for schools (The University of Adelaide 2021), with a specific focus on supporting students from low socioeconomics schools (SES), remote and regional schools, in addition to schools with high Indigenous student enrollments.

- Digital Technologies in Focus (DTiF) delivered by ACARA—A team of DT educators working with school leaders of around 160 disadvantaged schools to drive change in their schools through face-to-face workshops, webinars, and online mentoring.
- STEM Professionals in Schools by CSIRO—A program brokering flexible partnerships between STEM professionals and schools.
- Digital Literacy School Grants—A funding scheme for schools to pitch and implement their own innovative DT projects.
- Australian DT Challenges by the Australian Computing Academy (ACA) at the University of Sydney—An online series of structured, progressive teaching and learning activities and challenges for years 3 to 8 students and professional learning workshops for teachers, coding activities, and challenges for years 3 to 12 students.
- digIT summer schools (Australian Mathematics Trust)—Intensive camps and mentoring to target students at risk of not benefiting from the AC: DT to engage them in ICT and future careers.
- A Digital Technologies Hub (Education Services Australia 2021) was launched in 2016 that continues to operate. The Hub is an online learning resource that supports the implementation of the AC: DT with support for schools, families, and students.

The majority of this funding came to an end in 2020, with new funding initiatives for STEM in early learning and schools announced in the 2020–2021 budget (Australian Curriculum 2020) ($40.6 million). In relation to DT, this initiative supports artificial intelligence in schools (e.g., through lending library equipment from the University of Adelaide and the expansion of the Digital Technologies Hub), and the extension of the CSIRO STEM Professionals in Schools program.

Australia has had a very collaborative stakeholder ecosystem working with a common goal to support teachers and students with the AC: DT. This has been encouraged through linkages in the NISA initiatives, national and local DT association conferences and events, and networking fostered through Google Australia's annual Partner Summit in which they bring industry, education, and academic institutions working in the DT space across New Zealand and Australia together to attend workshops, hear updates, and build networks.

There were 30,585 primary and 24,976 secondary (ACARA 2021j) pre-service teachers enrolled to study teaching in 2019. Early research found pre-service teachers were ill prepared to teach CT and required pedagogical strategies, experience with relevant technologies, and a better understanding of how computers work and what CT means (Bower and Falkner 2015). Some pre-service teacher programs include DT but (as is the case in New Zealand) this is not explicitly mandated yet, nor has there been a consistent national approach to embedding DT within teacher education programs. In recognizing that a pipeline of new teachers entering classrooms also needed support, initiatives such as Google's CS4HS and CSER (Computer Science Education Research Group) DT Education program provided support to interested universities through workshops and resources.

INTENTIONS

NEW ZEALAND

As described earlier, the part of the NZC relating to digital technologies is two series of "progress outcomes" (POs) (New Zealand Ministry of Education 2021c) that cover "designing and developing digital outcomes" (DDDO) and "computational thinking for digital technologies" (CTDT), respectively. For each of these two series, the final three POs correspond to the three levels of the NCEA qualification, which students typically complete in years 11 to 13 of their schooling.

The series of POs that correspond to computer science are the eight POs for "computational thinking for digital technologies" (CTDT). The (POs) are essentially cumulative, and the PO corresponding to what typical year 10 students are likely to encounter corresponds to what most students should get to during their secondary school education (typically by year 10). In CTDT, this is PO 5, which is:

In authentic contexts and taking account of end-users, students independently decompose problems into algorithms. They use these algorithms to create programs with inputs, outputs, sequence, selection using comparative and logical operators and variables of different data types, and iteration. They determine when to use different types of control structures.

Students document their programs, using an organised approach for testing and debugging. They understand how computers store more complex types of data using binary digits, and they develop programs considering human-computer interaction (HCI) heuristics.

CTDT POs 1 to 5 gradually introduce sequence, input, selection, and iteration so that by PO 5 students are accessing the full power of computation. In addition, evaluating interfaces (and a focus on the end-user) and the representation of data are introduced, as well as the idea that there can be multiple algorithms to solve a particular problem.

There are six POs for DDDO that are focused on digital tools to design a range of "outcomes" (which could be a 3D-printed object, a website, poster, video, or anything where skill with a digital tool is central to the design and production process). There is a focus on the stakeholders for which the outcome is being produced, including ethical and social considerations (such as privacy). The outcomes expect increasing independence from students to select and apply tools. The DDDO PO corresponding to what most secondary students are expected to encounter is PO3 (see New Zealand Ministry of Education 2021c).

Beyond year 10, students typically start to specialize in subjects, and the curriculum offers the opportunity to delve deeper into computing topics. This structure is discussed in the following section under "Examination/ Certification."

AUSTRALIA

The AC: DT presents content descriptions across bands from F–10 (F–2, 3–4, 5–6, 7–8, and 9–10). The content advises what students should be learning, and achievement standards describe what students should achieve by the end of that year level band, which we discuss next (ACARA 2021k). As mentioned previously, the AC: DT includes two overarching key strands similar to New Zealand's DDDO and CTDT technological areas (see figure 5.1). Knowledge and understanding contains ten content descriptions that address underpinning disciplinary knowledge of information systems, including digital systems—covering the components of digital systems: hardware, software, and networks and their use—and representation of data, including how data are represented and structured symbolically. To demonstrate the growth in complexity of knowledge and understanding required,

By the end of Year 2 [the first band], students identify how common digital systems (hardware and software) are used to meet specific purposes. They use digital systems to represent simple patterns in data in different ways (ACARA 2021k).

Disciplinary knowledge deepens and "by the end of Year 10 students explain the control and management of networked digital systems and the security implications of the interaction between hardware, software and users. They explain simple data compression, and why content data are separated from presentation."

Processes and production skills contains thirty-three content descriptions relating to the development of skills to create digital solutions to problems and opportunities. These specifically address collecting, managing, and analyzing data and the creation of digital solutions by investigating and defining, generating, and designing, producing and implementing, evaluating, and collaborating and managing projects. Implicitly, the curriculum requires the application of CT skills. In this strand, by the end of year 2,

students design solutions to simple problems using a sequence of steps and decisions. They collect familiar data and display them to convey meaning. They create and organise ideas and information using information systems, and share information in safe online environments. (ACARA 2021k)

As programming is not explicitly mentioned in F–2, understanding of algorithms can be achieved with students engaging in algorithms with unplugged (e.g., the familiar "jam sandwich" activity), or plugged-in activities (e.g., a junior programming app or robotic device). From years 3 to 4 the curriculum prescribes the use of visual programming environments, involving branching (decisions) and user input, with iteration being introduced from years 5 to 6. From years 7 to 8 students are required to use general-purpose programming languages, using branching, iteration, and functions. In years 9 to 10, students move on to object-oriented programming languages by applying selected algorithms and data structures (this is in contrast to New Zealand, where object-oriented programming is not expected until year 13). This strand also includes content descriptions relating to the collection and visualization of data, intersecting with mathematics. As we move along the bands, students' digital solutions grow in complexity and by the end of year 10,

students plan and manage digital projects using an iterative approach. They define and decompose complex problems in terms of functional and non-functional requirements. Students design and evaluate user experiences and algorithms. They design and implement modular programs, including an object-oriented program, using algorithms and data structures involving modular functions that reflect the relationships of real-world data and data entities. They take account of privacy and security requirements when selecting and validating data. Students test and predict results and implement digital solutions. They evaluate information systems and their solutions in terms of risk, sustainability and potential for innovation and enterprise. They share and collaborate online, establishing protocols for the use, transmission and maintenance of data and projects. (ACARA 2021k)

A number of key concepts and key elements of CT underpin the AC: DT, providing a framework for knowledge and practice and for approaching problems. ACARA identifies these as: abstraction, data collection, data representation, data interpretation, specification, algorithms, implementation, digital systems, interactions, and impacts.

EXAMINATION/CERTIFICATION

NEW ZEALAND

Formal examination for most New Zealand students starts at year 11, and the majority of students will study for NCEA (New Zealand Qualifications Authority 2021), which is a national qualification. The NCEA qualification in DT is assessed primarily through a collection of "achievement standards." A single subject for a year is typically assessed by combining several of these standards. For example, in DT there are (at the time of writing) eleven achievement standards available at "level 1" (typically year 11 students), covering topics such as "develop a proposal for a digital outcome," "develop a design for a digital outcome," "develop a computer program," and "demonstrate understanding of searching and sorting algorithms." These standards are worth between three and six credits (one credit corresponds to approximately ten hours of student work), and a typical school course would be assessed using around eighteen to twenty credits. This means that a DT course in a school would usually access fewer than half of the standards, and while there are some rules about combining them, this gives a lot of flexibility to the school to use combinations of these standards to offer courses that focus on areas as varied as web design, programming, and computer science, or designing and developing 3D-printed objects. Students within a class might even focus on different tools and technology, so the curriculum provides a lot of flexibility to adjust for local interests. For employers and tertiary organizations looking at qualifications, they will typically focus on the combination of achievement standards that a student has, rather than the name of the courses they have taken. For example, one student may have done a lot of programming and computer science in their DT classes, while another may have had a strong focus on designing and developing products in particular media.

This vast range of choices for students can be problematic since so many pathways are possible, and there can be a lot of decisions for a student to make. A "review of achievement standards" (RAS) was started in 2018, with a view to having revised standards fully implemented by 2025 (New Zealand Ministry of Education 2021f). This includes moving to fewer, larger standards with a simpler structure and clearer pathways for learners. The curriculum (i.e., the progress outcomes described earlier) is not changing, but the way it is assessed will be changed, which is likely to be reflected in less fragmentation of students' coverage of the curriculum.

AUSTRALIA

Australia has final year examinations (year 11 and 12), but this is not captured in the Australian Curriculum, as it is up to the states/territories to develop and assess. States and territories vary in what year 11 and 12 subjects offer, and schools can vary within states and territories. For example, Victoria offers "algorithmics" (algorithmic problem-solving) and "applied computing" (involving data analytics and software engineering) under their digital technologies offerings with some also in a vocational education and training (VET) pathway (Victorian Curriculum and Assessment Authority 2021). Western Australia offers computer science, covering databases, systems, programming, and networks, as well as applied information technology, including the application of hardware and software to create solutions (Government of Western Australia 2021).

Queensland offers digital solutions, covering algorithms, computer languages, user interfaces, and generating solutions as well as ICT covering hardware, software, and ICT in society (Queensland Curriculum and Assessment Authority 2021). The differences in language, depth of CS disciplinary content, and offerings across the eight Australian states and territories make it difficult to ensure that students have access to the same CS opportunities following their F–10 curriculum and also make it difficult to measure the CS pipeline.

In terms of uptake of DT in schools in senior secondary years, the Australian government Department of Education Skills and Employment (DESE) data, available on the ACARA website, shows that in 2019, 31.5 percent (39.1 percent males, 24.3 percent females) elected to study ICT and design and technology in year 12 (figure 5.4). These are two distinct subject areas and reporting them together is problematic for monitoring DT uptake. Despite the introduction of the AC: DT and supporting initiatives, we are yet to see this translate to students' uptake of DT in senior secondary years. This could be due in part to the lack of years 11 to 12 DT in the AC, causing inconsistencies in the way states and territories implement within these years.

The National Assessment Program—Literacy and Numeracy (NAPLAN) is an annual national assessment for all students in years 3, 5, 7, and 9. All

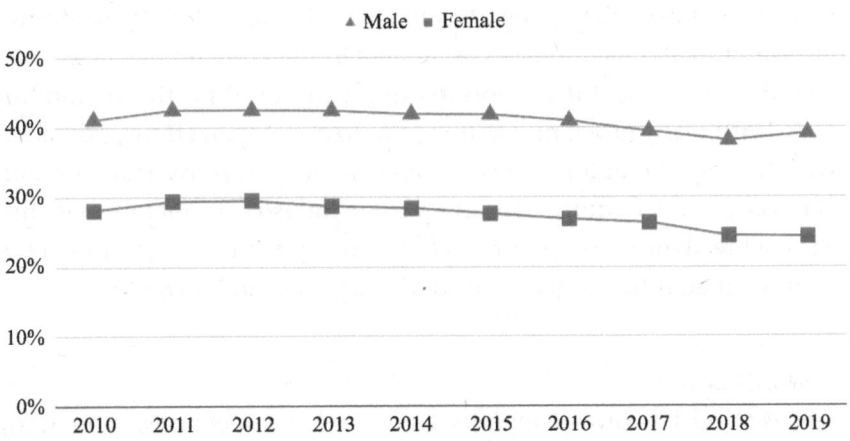

5.4 Percentage of students enrolled in information and communication technology and design and technology by year (based on ACARA 2021l).

students in these year levels are expected to participate in tests in reading, writing, language conventions (spelling, grammar, and punctuation), and numeracy. Additionally, the National Assessment Program—Information and Communication Technology Literacy (NAP–ICTL), which commenced in 2005, measures achievement in ICT literacy. It involves the assessment of samples from students in year 6 and year 10 in schools via purpose-built software applications carefully designed to reflect "real-world" ICT contexts familiar to students (e.g., using the Internet for research and designing a digital poster). In 2017, the NAP–ICT Literacy Assessment Framework was revised to describe and represent its relationship to the AC: ICT Capability and AC: DT (see page 4 in ACARA 2018). DT links include those relating to managing and operating digital systems; accessing, evaluating, and analyzing data and information; creating with digital systems; collaborating and communicating; and using digital systems appropriately. The results of the 2017 NAP–ICTL report showed that, in comparison to the previous test in 2014, the results were stable and female students were outperforming male students (ACARA 2018).

EXTRACURRICULAR ACTIVITIES

Extracurricular activities such as coding clubs and robotics competitions are important to develop a pathway for students who want to pursue their interest in DT beyond the opportunities in their school. Prior to the introduction of digital technologies as an area in the curriculum, these were often the only outlet for students to explore these skills with like-minded peers. With many programs existing prior to curriculum efforts, primarily with the aim of building a diverse pipeline of prospective students and addressing gender equity concerns within industry, in many cases this provided readymade support for schools to explore this new curriculum effort, with an inherent focus on equity, diversity, and inclusion.

NEW ZEALAND

New Zealand has several robotics groups and competitions; most were operating from before the curriculum was introduced and are still thriving. These include Robocup Junior (Robocup Junior New Zealand 2021), which is based on challenges in specific contexts, such as playing a simple sports game; FIRST Lego League (FIRST New Zealand 2021); and Vex Robotics (Kiwibots 2021).

Several other national competitions also operate. The "123Tech" challenge (ITP NZ 2021) provides a range of competitions for students of all ages and is run by IT Professionals New Zealand (a national industry organization). Junior students have challenges involving CS Unplugged activities (Bell and Vahrenhold 2018), while senior students compete around designing applications. An annual final event with prizes provides an incentive for students to engage with digital technologies. New Zealand also has programming competitions, with the New Zealand Olympiad in Informatics (New Zealand Olympiad in Informatics 2021) as a springboard for school students to compete in the International Olympiad in Informatics (IOI). In an addition, an event called "Programming Challenge for Girls" (PC4G) has been run in main centers, giving year 10 girls an opportunity to learn programming and demonstrate their skill; this has led to the students finding their passion in programming and carrying on a digital technologies pathway. Another international competition that is gaining traction in NZ is the Bebras challenge.

AUSTRALIA

There are a growing number of extracurricular opportunities for students in Australia, including many that were established prior to the AC: DT with a focus on increasing equity and participating in computing, and a number of new programs supporting curriculum changes. Broadly, there are 496 activities for "robotics" and "technology and design" published on the STAR portal for students at the time of writing (Australian Government 2021), and sixteen competitions and nine challenges are listed on the Digital Technologies Hub. Code Clubs Australia, Bebras Australia, the Hour of Code, TechGirls Movement, FIRST Robotics, and NCSS Challenge have all been popular national activities. A number of extracurricular providers have started to map their programs to the AC: DT, which has supported schools to engage and align their extracurricular activities both inside and outside of the classroom with the intended curriculum. We discuss some examples in the following paragraphs.

The Grok Learning team have been running coding competitions for more than twelve years though the NCSS Challenge, a programming competition open to all school students and teachers. The NCSS Challenge is for school students from upper primary to all levels of high school with content aligned to the AC: DT for years 5 to 10, and streams to match a variety of student abilities. NCSS operates on a small fee per student that provides access to online coding tutorials, access to instructor support, and entry to the competition; it includes free access for teachers (pre-service and in-service). Participants compete against students from Australia and New Zealand and earn points to climb the leaderboard.

FIRST Australia is a program based on the original successful international robotics program (FIRST Australia 2021). It provides school students with a suite of competitions targeting different age groups from four to sixteen years of age, aimed at inspiring a passion for STEM through robotics challenges and competitions.

The annual "Search for the Next Tech Girl Superhero" competition aims to increase girls' participation in technology careers and engages up to one thousand Australia and New Zealand students per year with a total of over ten thousand girls since 2014. This competition, in which girls design, pitch, and prototype an app solution, is for ages between seven and seventeen. It connects girls to mentors as they work in teams. Most teams meet at schools and are facilitated by school teachers. In some cases, parents and other community leaders step up to facilitate teams. While mentorship competitions are having a positive impact, they do rely on volunteer mentors with the confidence to support a team; however, the program has also mapped to the AC: DT supporting teachers to integrate participation with school programs. Curious Minds, funded through the Australian government, is an example of a broader STEM program that operates with a similar mentorship process in which girls in years 9 to 10 attend two camps and work with their mentors on a project (Australian Science Innovations 2021).

Bebras is an international initiative, with some sixty countries and over 2.9 million students worldwide, that promotes informatics and computational thinking among students. Bebras, meaning "beaver" in Lithuanian, was started in 2004 by Professor Valentina Dagiene. Bebras Australia launched in 2014 and is now administered through CSIRO Digital Careers, taking place March and August–September each year. Students can participate individually or in teams (up to four people) on any day during a two-week period, providing flexibility for school schedules. The challenge is broken up into five age groups across years 3 to 12, with forty-five to sixty minutes for students to complete. Each challenge features fifteen CT questions to solve.

There are not many student-facing Aboriginal and Torres Strait Islander programs specifically in DT. However, one recent example is that in 2014, Indigital launched Australia's first Indigenous edu-tech company with a focus on training in digital technologies with a cultural lens (Indigital 2019). Indigital began "Indigital Schools" in early 2020, a student-facing program that enables Indigenous and non-Indigenous primary and high school students to "connect with and learn from Indigenous Elders about cultural knowledge, history, and language while learning digital skills" in technologies like augmented reality, animation, and coding.

MEDIA AND TEACHING RESOURCES

To deliver the new curriculum material effectively, good teaching resources are needed for teachers to use in the classroom. Most resources supporting the changes in Australia and New Zealand are online, although some books have been published that match the Australian DT curriculum.

In both countries, schools are not given required texts to use, so teachers are free to use whatever resources they consider suitable. This also means that the costs of such resources come from school budgets, and buying class sets or subscriptions to online systems can be a challenge for some schools where resources are spread thin, so free online resources are particularly welcome.

NEW ZEALAND

The official online repository of resources to support New Zealand schools is Te Kete Ipurangi (TKI), which means "the online knowledge basket" (New Zealand Ministry of Education 2021g). It is an initiative of the Ministry of Education and includes the curricula, guidance for teachers, and many teaching resources. There is an area dedicated to the technology learning area (New Zealand Ministry of Education 2021h) and within that, support for DT. Everything posted on the site goes through a quality assurance process.

When computer science topics appeared in NCEA standards in 2011, there was little supporting media for teaching these at the right level. For example, a teacher might be having to cover "formal languages" as a subject for the first time. It was unlikely that they had access to books on the subject, and if they searched online, they would either find lecture notes that quickly went to a depth way beyond the expectation of the standard or might be directed to explanations on formal *natural* language, such as business letters and sermons (Munasinghe, Bell, and Robins 2021). To address this issue, the "Computer Science Field Guide" (CSFG) was created at the University of Canterbury as an open educational resource that could be quickly updated to reflect the needs of schools (Bell, Duncan, Jarman, and Newton 2014). The title comes from Peter Denning, who envisaged a kind of handbook that school students could use for learning about computer science, particularly topics other than learning to program (for which many resources are available). The chapters in the CSFG primarily correspond to topics that appear in the New Zealand NCEA standards.

Other online resources to specifically support aspects of the curriculum have also appeared, including CodeAvengers (CodeAvengers 2021) and "STEM Online NZ" (University of Auckland 2021), which offer online lessons to match particular DT topics.

For teaching programming, there are many resources available internationally, although they often do not fit well with New Zealand culture (including simple issues like summer starting in December, the color of school buses, or the use of idioms), which means New Zealand students would not see themselves in the situations depicted in such material. Thus, New Zealand–based initiatives have also arisen around students learning programming, such as CodeAvengers (CodeAvengers 2021) and Gamefroot (Gamefroot 2021).

The DTTA teachers' association, which was created to support teachers involved in the DT curriculum change, was also an important source of resources. The association runs an online forum where teachers share resources, including those that they have generated themselves. The archives of this list became a valuable, even if difficult to search, repository of information. More recently, the DTTA supported the creation of the "DTHM4Kaiako" site (DTHM for kaiako 2021), which is intended to capture and index popular resources. The DTTA also piloted a lending library where teachers can temporarily borrow class sets of devices such as robots or micro:bits to try them out for a term before buying their own sets. In addition to the DTTA, TENZ (Technology Education New Zealand) is a professional association that supports teachers in the whole technology area.

AUSTRALIA

Prior to the launch of the AC: DT, a review of F–10 DT resources (Falkner and Vivian 2015) revealed that limited resources existed to support teachers in implementing the F–10 AC: DT. Many teachers have utilized familiar CS education software, digital technology, and lesson plans such as CS Unplugged (Bell and Vahrenhold 2018), Python tutorials, Code.org, Bee-Bots, and Scratch by MIT. As new technologies have emerged to support younger audiences, schools have embraced these; however, many schools rediscovered technology such as Bee-Bots sitting in their cupboards from prior to the introduction of AC: DT. Australia is a country with extremely remote schools and vast differences in resource availability. Teachers value initiatives and resources that present "unplugged" approaches for teaching CT and CS in situations in which computers and DT equipment are not readily accessible.

The Digital Technologies Hub (Education Services Australia 2021), funded through the Australian government, is a national platform for resources and support for the AC: DT learning resources and services for teachers, students, school leaders, and parents and is known as the place to go for DT resources and support with programs and curriculum in schools. The Hub also curates high-quality events and activities offered by education jurisdictions, industry, and other providers. Over the years, the Hub has expanded to provide additional support in specific areas as AC implementation and teacher needs have evolved, for example, assessment resources, inclusive DT education, and emerging topics such as AI. The site includes support for families and students. Having support for families is particularly important and shows recognition of the important role that families play in shaping students' attitudes as well as influencing what schools offer.

With NISA funding, the Computer Science Education Research Group (CSER) at The University of Adelaide in 2016 expanded their initial industry-funded F–6 DT MOOC with 5,727 teacher enrollments to an expanded ecosystem of program support for teachers in schools (Falkner, Vivian, and Williams 2018). The program included a series of community-centric MOOCs for teachers that break down the AC: DT curriculum, concepts, and class activities with connections to high-quality programs and resources; a free national digital technology lending library enabling schools access to DT equipment; and free professional learning workshops delivered by project officers across the country (The University of Adelaide 2021). In partnership with Google Australia, CSER launched a PL-in-a-Box program to support grassroots uptake of our MOOCs with readymade workshop packs. A team of project officers driving MOOC engagement and school access to free DT equipment, with priority to target schools (rural, disadvantaged, and schools with high Aboriginal and Torres Strait Islander enrollments), has been critical to engaging schools that need support the most and has been demonstrated with significant increases in target school participation, from 14.7 percent with a MOOC alone to 30.4 percent with additional support. Since 2014, the program has evolved to include eight MOOCs covering cutting-edge areas, such as cybersecurity and artificial intelligence, with industry and government support. In a government-commissioned evaluation of all NISA early learning and school STEM initiatives (Dandolo Partners 2020), the program was cited as a "stand-out NISA initiative" and "stakeholders interviewed consistently report positive feedback about MOOCs, in particular its ability to translate digital technologies concepts . . . MOOCs is widely credited as making a significant difference in teachers' understanding of digital technologies." The CSER MOOCs and PL-in-a-Box model, in combination with a national Hub of resources, has become a blueprint for the rollout of a government-funded national program for mathematics and numeracy education. During the NISA 2016–2020 period, CSER worked with over 38,000 teachers across Australia, including 11,300 teachers from priority schools. Project officer and lending library funding came to an end in December 2020, with the exception of an extension for a number of AR and VR kits until 2022 under an AI in schools initiative. The MOOCs and associated resources are creative commons open source and remain freely available.

The Australian Computing Academy (ACA) at the University of Sydney was formed with Australian government NISA funding to develop a series of online free coding challenges and resources for students in years 3 to 8 (Australian Computing Academy 2021) extending the work of the NCSS challenges by Grok Learning. ACA used the existing Grok Learning platform, an education start-up from the University of Sydney that operated a subscription model for coding challenges. After government funding ended, ACA and Grok Learning merged to form the Grok Academy, a not-for-profit organization providing online DT resources. As ACA's DT Challenges were funded by the Australian government until the end of 2020, from 2021 onward, the DT coding challenges and any new content developed require a paid Grok Academy subscription. The ACA's unplugged activities that were developed with government funding (e.g., DT@home, Unpacking the Curriculum guide, and webinar recordings) remain free on their website. With recent interest and focus from the government, ACA has also developed a Schools Cyber Security Challenges series for years 5–12 in partnership with AustCyber and Australian industry partners. This is currently free for Australian schools.

As the majority of government funding has come to an end for a number of NISA programs supporting AC: DT rollout, those projects are now navigating how to move forward in a sustainable way so they can continue to support schools with free resources and programs.

Besides government funding, there are some other broader STEM initiatives that support AC: DT in schools, such as Refraction Media's Careers with STEM platform (Refraction Media 2021) that includes digital magazines and articles covering careers with code and special editions such as cybersecurity and data science. They are funded by industry, print subscriptions, and an advertizing model. These magazines have been popular for advocating career awareness and are sent to schools across Australia, reaching 750,000 students annually in person, print, and online. With the success of Careers with STEM, they have expanded to new international audiences, including New Zealand and the United States of America.

While some of the NISA initiatives have had a focus on prioritizing support for schools with high Aboriginal and Torres Strait Islander student enrollments and some resource development, such as ACARA's DTiF classroom ideas and worked examples (ACARA 2021m), many of the current initiatives teach primarily Western views of DT, and more work is required to support teachers to integrate DT with Aboriginal and Torres

Strait Islander peoples, cultures, and histories. Examples of programs leading in this space include the Stronger Smarter Institute (SSI) partners with
schools across Australia to embrace their pedagogical framework centered
around honoring a positive sense of Aboriginal and Torres Strait Islander
cultural identity and positive community leadership (Stronger Smarter
Institute 2017). A number of new elaborations have been developed in
the proposed revised curriculum to further support the cross-curriculum
priorities. The AC: DT elaborations provide a reference point for ideas,
but investment in teacher professional learning and resources is required.

There are a number of computing associations across Australia that provide networking and professional development for teachers, including the
national Australian Council for Computers in Education (ACCE) and local
state and territory technology associations for teachers (e.g., ICTENSW,
TASITE, EdTechSA, INTEACT, QSITE, and DLTV). Prior to the AC: DT these
associations focused on using ICT; however, with the introduction of new
computing curriculum, they have expanded their focus to include DT.

CURRICULUM ISSUES

Introducing a new curriculum in school systems where many of the staff
have not had the subject as part of their own education is always a challenge. As well as the usual confusion around the term "digital technologies" being about "using ICT" rather than developing systems, there is
the issue of fitting the material into an already crowded curriculum.

Of interest in this situation is the "enacted" curriculum—what actually
happens in schools compared with the intended curriculum. For example, in Australia a pilot study (Falkner et al. 2019b) found that the most
enacted part of the DT curriculum was the design process (reported by
86 percent of teachers); 79 percent reported teaching programming skills
and concepts, 57 percent had taught data representation (e.g., digital
data, binary), and only 29 percent enacted DT ethics. On the other hand,
topics that were *not* specified in the curriculum were also enacted, including robotics (79 percent). These differences could be explained by the
availability of professional learning as well as an emphasis on "concepts
that teachers find are useful or relevant to include or link to within the
intended curriculum" (Falkner et al. 2019b). The low level of engagement

with ethics could be improved by offering more structured support for teachers to consider ethical issues through appropriate lenses and at a level of technical depth. Any move in this direction should also be wary of detracting from the technical aspects of a topic, as teachers who are still developing their understanding of advanced topics may be tempted to spend *too much* time looking at ethics instead of the technical considerations of a topic. As an example of keeping this balance, there would be value in looking at specific algorithms or data encryption techniques and their implications for individuals and law enforcement rather than just having a general discussion on whether technology is good or bad.

In New Zealand, some data is available through two Education Review Office (ERO) reports. The first is a July 2019 report titled "It's Early Days for the New Digital Technologies Curriculum" (New Zealand Education Review Office 2021a), which used data collected in September 2018, along with case studies in early 2019. This report notes that "the assumption that teachers are digitally fluent as they participate in the support for the DT curriculum content does not hold" and that "only seven percent [of schools] said their teachers sufficiently understood the DT curriculum content and its place in the NZC and had enough knowledge and skills to implement the DT curriculum content." The second ERO report (New Zealand Education Review Office 2021b), from January 2020, focuses on three stages that schools were at: "On Your Marks" (e.g., very early stages of unpacking the DT curriculum content), "Get Set" (e.g., DT curriculum content most likely delivered in a club, specialist classes or subject specific setting), and "Go!" (e.g., in the process of rolling out DT curriculum content across all year levels). The report does not give statistics on how many schools were in these different stages, but it provides examples of what these stages looked like across a range of criteria, such as school leadership engagement, curriculum integration with other learning areas, and professional development content.

New Zealand and Australia have adopted two different approaches to an inclusive curriculum for First Nations peoples. Australia has included Aboriginal and Torres Strait Islander peoples, cultures, and histories throughout the AC, whereas New Zealand provides a model for how a country can operate a school system under two curricula: the "New Zealand Curriculum" (NZC) for English medium instruction, and Te Marautanga o Aotearoa (TMoA) for Māori medium instruction (New Zealand Ministry of Education 2021a). New Zealand has also invested in programs supporting Māori education with DT, such as the bicultural Kia Takatū ā-Matihiko program, providing teacher support and resources in both English and te reo Māori for DT. Australia has provided targeted support to schools with high Torres Strait Islander enrollments.

While the changes to the curriculum are important, considerable resourcing is needed to support teachers and schools to make the change, and while a good start has been made in both countries, there is still a lot of work to be done to have the subject fully available as intended for students and to upskill the vast number of teachers. Additionally, support for new emerging areas of DT, such as cybersecurity and artificial intelligence, bring new opportunities and will require professional learning and resourcing to evolve.

CONCLUSION

Both New Zealand and Australia have strong curricula for introducing all school students to digital technologies as a discipline, including elements of computational thinking, particularly computer programming. Although good progress has been made implementing this in schools, there is a long way to go as the subject becomes normalized in the curriculum, it is understood properly by school management, and teachers develop the professional skills to deliver the curriculum as intended. Additionally, Australia has reached a point where the AC: DT is now undergoing a curriculum review, which will determine what the intended DT will look like for classrooms for the near future and possibly any further strategic directions for implementation. Following initial bursts of funding and efforts to support new computing curriculum rollout, there is a challenge to establish sustainable models of accessible and equitable support for schools and for teachers to engage with the intended curriculum. Ongoing research will play a critical role in monitoring the intended curriculum as prescribed by New Zealand and Australia, and tracking the enacted and assessed curriculum that happens in the classroom will be important in understanding how teachers are embracing these curriculum changes many years after its implementation. Work by Falkner et al. (2019a) to develop the MEasuring TeacheR Enacted Computing Curriculum

(METRECC) instrument to measure the enacted and intended curriculum across several countries may provide a tool for ongoing measurement within and between the two countries.

REFERENCES

ACARA. 2018. "NAP Sample Assessment ICT Literacy: Years 6 and 10." https://www .nap.edu.au/docs/default-source/default-document-library/2017napictlreport_final .pdf.

ACARA. 2019. "International Comparative Study: The Australian Curriculum and the New Zealand Curriculum." https://www.australiancurriculum.edu.au/resources -and-publications/publications/program-of-research-2017-2020.

ACARA. 2021a. "National Report on Schooling in Australia 2013: School Structures." https://www.acara.edu.au/reporting/national-report-on-schooling-in-australia /national-report-on-schooling-in-australia-2013/schools-and-schooling/school -structures.

ACARA. 2021c. "Australian Curriculum: Technologies: Structure." www.australiancur-riculum.edu.au/f-10-curriculum/technologies/structure/

ACARA. 2021d. "Australian Curriculum: Aboriginal and Torres Strait Islander Histo-ries and Cultures." https://www.australiancurriculum.edu.au/f-10-curriculum/cross -curriculum-priorities/aboriginal-and-torres-strait-islander-histories-and-cultures/.

ACARA. 2021e. "Australian Curriculum: Foundation—Year 10 Curriculum." https:// www.acara.edu.au/curriculum/foundation-year-10.

ACARA. 2021f. "Australian Curriculum: Information and Communication Technol-ogy (ICT) Capability." https://www.australiancurriculum.edu.au/f-10-curriculum /general-capabilities/information-and-communication-technology-ict-capability/.

ACARA. 2021g. "Australian Curriculum: Development of the Australian Curriculum." https://www.acara.edu.au/curriculum/history-of-the-australian-curriculum/develop ment-of-australian-curriculum.

ACARA. 2021h. "Australian Curriculum: Review of the Australian Curriculum." https://www.acara.edu.au/curriculum/curriculum-review.

ACARA. 2021i. "National Report on Schooling Data Portal." https://www.acara .edu.au/reporting/national-report-on-schooling-in-australia/national-report-on -schooling-in-australia-data-portal.

ACARA. 2021j. "Report: Domestic Enrolments in All Teacher Education Courses by Field of Education, Australia, 2019." https://www.acara.edu.au/reporting/national -report-on-schooling-in-australia/national-report-on-schooling-in-australia-data -portal/teacher-education#view1.

ACARA. 2021k. "Australian Curriculum: Digital Technologies Curriculum." https:// www.australiancurriculum.edu.au/f-10-curriculum/technologies/digital-technologies/.

ACARA. 2021l. "National Report on Schooling in Australia Data Portal Year 12 Subject Enrolments." https://www.acara.edu.au/reporting/national-report-on-schooling-in-australia/national-report-on-schooling-in-australia-data-portal/year-12-subject-enrolments#view2.

ACARA. 2021m. "Aboriginal and Torres Strait Islander Connections to Digital Technologies." https://www.australiancurriculum.edu.au/media/6641/classroom-ideas-f-10-aboriginal-and-torres-strait-islander-connections-to-digital-technologies.pdf.

Australian Computing Academy. 2021. https://aca.edu.au/.

Australian Curriculum. 2020. "Support for Science, Technology, Engineering and Mathematics (STEM)." https://www.dese.gov.au/australian-curriculum/support-science-technology-engineering-and-mathematics-stem.

Australian Government. 2014. "Industry Innovation and Competitiveness Agenda: An Action Plan for a Stronger Australia, Department of the Prime Minister and Cabinet." https://pmc.gov.au/resource-centre/domestic-policy/industry-innovation-and-competitiveness-agenda-report-action-plan-stronger-australia.

Australian Government. 2021. "STARportal." https://starportal.edu.au.

Australian Government Department of Industry, Science, Energy and Resources. 2015. "National Innovation and Science Agenda Report." https://www.industry.gov.au/data-and-publications/national-innovation-and-science-agenda-report.

Australian Science Innovations. 2021. "What Is Curious Minds?" https://www.asi.edu.au/programs/curious-minds/what-is-curious-minds/.

Bell, Tim. 2014. "Establishing a Nationwide CS Curriculum in New Zealand High Schools." *Communications of the ACM* 57, no. 2: 28–30. https://doi.org/10.1145/2556937.

Bell, Tim, Peter Andreae, and Lynn Lambert. 2010. "Computer Science in New Zealand High Schools." In *ACE 2010: Proceedings of the Twelfth Australasian Conference on Computing Education* (Vol. 103), edited by Tony Clear and John Hamer, 15–22. Darlinghurst, NSW, Australia: Australian Computer Society.

Bell, Tim, Peter Andreae, and Anthony Robins. 2014. "A Case Study of the Introduction of Computer Science in NZ Schools." *ACM Transactions on Computing Education* 14, no. 2 (June): 10:1–10:31. http://doi.acm.org/10.1145/2602485.

Bell, Tim, Caitlin Duncan, Sam Jarman, and Heidi Newton. 2014. "Presenting Computer Science Concepts to High School Students." *Olympiads in Informatics* 8: 3–19.

Bell, Tim, and Jan Vahrenhold. 2018. "CS Unplugged—How Is It Used, and Does It Work?" In *Adventures Between Lower Bounds and Higher Altitudes: Essays Dedicated to Juraj Hromkovič on the Occasion of His 60th Birthday (Vol. 11011)*, edited by Hans-Joachim Böckenhauer, Dennis Komm, and Walter Unger, 497–521. Cham: Springer.

Blum, Lenore, and Thomas J. Cortina. 2007. "CS4HS: An Outreach Program for High School CS Teachers." In *Proceedings of the 38th SIGCSE Technical Symposium on*

Computer Science Education, edited by Ingrid Russell, Susan M. Haller, J. D. Dougherty, and Susan H. Rodger, 19–23. Covington, KY: ACM. http://dblp.uni-trier.de/db /conf/sigcse/sigcse2007.html#BlumC07

Bower, Matt, and Katrina Falkner. 2015. "Computational Thinking, the Notional Machine, Pre-Service Teachers, and Research Opportunities." In *Proceedings of the 17th Australasian Computing Education Conference (ACE 2015) Vol. 160*, edited by Daryl D'Souza and Katrina Falkner, 37–46. Sydney: ACS.

Code Club. 2021. "Code Club 4 Teachers." https://codeclub.nz/teachers.

CodeAvengers. 2021. https://www.codeavengers.com/.

Curious Minds. 2021. "Digital Technologies, Hangarau Matihiko." https://www .curiousminds.nz/actions/education/curriculum/digital-technologies-hangarau -matihiko/.

Dandolo Partners. 2020. "Evaluation of Early Learning and Schools Initiatives in the National Innovation and Science Agenda." https://www.dese.gov.au/download /4826/evaluation-early-learning-and-schools-initiatives-national-innovation-and -science-agenda/7196/document/pdf.

Digital Ignition. 2021. "Digital Ignition | Māpura Matihiko." https://www.digital ignition.co.nz/.

DTHM for kaiako. 2021. https://www.dthm4kaiako.ac.nz/.

Education Services Australia. 2019. "Alice Springs (Mparntwe) Education Declaration." https://www.dese.gov.au/indigenous-education/resources/alice-springs-mparntwe -education-declaration.

Education Services Australia. 2021. "Digital Technologies Hub." https://www .digitaltechnologieshub.edu.au/.

Falkner, Katrina, and Rebecca Vivian. 2015. "Coding across the Curriculum: Resource Review. Australian Government: Department of Education and Training." https://www.dese.gov.au/uncategorised/resources/resource-review-report-coding -across-curriculum.

Falkner, Katrina, Sue Sentence, Rebecca Vivian, Sarah Barksdale, Leonard Busuttil, Elizabeth Cole, Christine Liebe, Francesco Maiorana, Monica M. McGill, and Keith Quille. 2019a. "An International Study Piloting the MEasuring TeacheR Enacted Computing Curriculum (METRECC) Instrument." In *ITiCSE-WGR '19: Proceedings of the Working Group Reports on Innovation and Technology in Computer Science Education*, edited by Guido Rößling, Michail Giannakos, Bruce Scharlau, Roger McDermott, Arnold Pears, and Mihaela Sabin, 111–142. New York: ACM.

Falkner, Katrina, Sue Sentence, Rebecca Vivian, Sarah Barksdale, Leonard Busuttil, Elizabeth Cole, Christine Liebe, Francesco Maiorana, Monica M. McGill, and Keith Quille. 2019b. "An International Comparison of K-12 Computer Science Education Intended and Enacted Curricula." In *Proceedings of the 19th Koli Calling International*

Conference on Computing Education Research, edited by Petri Ihantola and Nick Falkner, 1–10. New York: ACM. https://doi.org/10.1145/3364510.3364517.

Falkner, Katrina, Rebecca Vivian, and Sally-Ann Williams. (2018). "An Ecosystem Approach to Teacher Professional Development within Computer Science." *Computer Science Education* 28, no. 4: 303–344. https://doi.org/10.1080/08993408.2018 .1522858.

Garry Falloon. 2020. "New Zealand's ICT-in-Education Development (1990–2018)." In *ICT in Education and Implications for the Belt and Road Initiative (Lecture Notes in Educational Technology),* edited by Looi, Chee-Kit, Hui Zhang, Yuan Gao, Longkai Wu, 133–148. Singapore: Springer. https://doi.org/10.1007/978-981-15-6157-3_8.

FIRST Australia. 2021. "Info for Schools." https://firstaustralia.org/get-involved/info -for-schools/.

FIRST New Zealand. 2021. https://www.firstnz.org/.

Fox-Turnbull, Wendy. 2019. "Implementing Digital Technology in the New Zealand Curriculum." *Australasian Journal of Technology Education,* article 5. https://doi.org /10.15663/ajte.v5i0.65.

Gamefroot. 2021. https://make.gamefroot.com/.

Google for Education. 2021. "Educator PD Grants." https://edu.google.com /computer-science/educator-grants/.

Government of Western Australia. 2021. "Syllabus and Support Materials: Technologies." https://senior-secondary.scsa.wa.edu.au/syllabus-and-support-materials /technologies.

Hubwieser, Peter, Michal Armoni, Torsten Brinda, Valentina Dagiene, Ira Diethelm, Michail N. Giannakos, Maria Knobelsdorf, Johannes Magenheim, Roland Mittermeir, and Sigrid Schubert. 2011. "Computer Science/Informatics in Secondary Education." In *Proceedings of the 16th Annual Conference Reports on Innovation and Technology in Computer Science Education—Working Group Reports* (Vol. Darmstadt), edited by Liz Adams and Justin Joseph Jurgens, 19–38. New York: ACM. https://doi .org/10.1145/2078856.2078859.

Indigital. 2019. "Indigital Schools." https://indigitalschools.com/.

ITP NZ. 2021. "Tahi Rua Toru Tech." https://123tech.nz/.

Kiwibots. 2021. https://www.kiwibots.co.nz/.

Ministerial Council on Education, Employment, Training, and Youth Affairs. 2008. "Melbourne Declaration on Educational Goals for Young Australians." https://docs .acara.edu.au/resources/national_declaration_on_the_educational_goals_for_young _australians.pdf.

Munasinghe, Bhagya, Tim Bell, and Anthony Robins. 2021. "Teachers' Understanding of Technical Terms in a Computational Thinking Curriculum." In *Proceedings of*

Australasian Computing Education Conference (ACE '21), edited by Claudia Szabo and Judy Sheard, 9 pages. New York: ACM. https://doi.org/10.1145/3441636.3442311.

Museum of New Zealand—Te Papa Tongarewa. 2021. "Raranga Matihiko | Weaving Digital Futures." https://www.tepapa.govt.nz/learn/for-educators/raranga-matihiko -weaving-digital-futures.

New Zealand Government. 2017. "$40m Digital Fluency Package." https://www .beehive.govt.nz/release/40m-digital-fluency-package.

New Zealand Olympiad in Informatics. 2021. http://www.nzoi.org.nz/.

New Zealand Qualifications Authority. 2021. "NCEA." https://www.nzqa.govt.nz /ncea/.

NSW Education Standards Authority. 2021. "Science and Technology K–6 Syllabus (2017)." https://educationstandards.nsw.edu.au/wps/portal/nesa/k-10/learning-areas /science/science-and-technology-k-6-new-syllabus.

NZ Education Review Office. 2021a. "It's Early Days for the New Digital Technologies Curriculum Content." https://www.ero.govt.nz/publications/its-early-days-for -the-new-digital-technologies-curriculum-content/.

NZ Education Review Office. 2021b. "On Your Marks . . . Get Set . . . Go! A Tale of Six Schools and the Digital Technologies Curriculum Content." https://www.ero .govt.nz/publications/on-your-marks-get-set-go-a-tale-of-six-schools-and-the-digital -technologies-curriculum-content-2/.

NZ Ministry of Education. 2021a. "The New Zealand Curriculum." https:// nzcurriculum.tki.org.nz/.

NZ Ministry of Education. 2021b. "The New Zealand Curriculum: Learning Areas." https://nzcurriculum.tki.org.nz/The-New-Zealand-Curriculum#collapsible9.

NZ Ministry of Education. 2021c. "The New Zealand Curriculum: Technology: Progress Outcomes." https://nzcurriculum.tki.org.nz/The-New-Zealand-Curriculum /Technology/Progress-outcomes.

NZ Ministry of Education. 2021d. "The New Zealand Curriculum: Technology in the New Zealand Curriculum (Revised)." https://nzcurriculum.tki.org.nz/content /download/168478/1244184/file/NZC-Technology%20in%20the%20New%20 Zealand%20Curriculum-Insert%20Web.pdf.

NZ Ministry of Education. 2021e. "The Kia Takatū ā-Matihiko | Digital Readiness Programme." https://kiatakatu.ac.nz/.

NZ Ministry of Education. 2021f. "NCEA Review." https://conversation.education .govt.nz/conversations/ncea-review/.

NZ Ministry of Education. 2021g. "TKI | Te Kete Ipurangi." https://www.tki.org.nz/.

NZ Ministry of Education. 2021h. "Technology Online." https://technology.tki.org .nz/.

NZ Ministry of Education. 2021i. "Technology Online." https://nzcurriculum.tki
.org.nz/var/tki-nzc/storage/images/media/images/tu-images/technology-structure
/1235735-3-eng-NZ/Technology-structure_preview.jpg.

NZ Ministry of Education. 2021j. "Strengthening Local Curriculum." https://
nzcurriculum.tki.org.nz/Strengthening-local-curriculum/About.

Queensland Curriculum and Assessment Authority. 2021. "Senior Secondary."
https://www.qcaa.qld.edu.au/senior.

Refraction Media. 2021. "Careers with STEM." https://careerswithstem.com.au/.

Robocup Junior New Zealand. 2021. https://sites.google.com/view/robocupjunior
newzealand.

Stronger Smarter Institute. 2017. "Implementing the Stronger Smarter Approach."
https://strongersmarter.com.au/wp-content/uploads/2020/07/Stronger-Smarter
-Approach-2017_final-2.pdf.

The University of Adelaide. 2021. "CSER Digital Technologies Education." https://
csermoocs.adelaide.edu.au/.

Thompson, David, and Tim Bell. 2013. "Adoption of New Computer Science High
School Standards by New Zealand Teachers." In *The 8th Workshop in Primary and
Secondary Computing Education (WiPSCE 2013)*, edited by Maria Knobelsdorf, Ralf
Romeike, and Michael E. Caspersen, 87–90. Aarhus, Denmark: ACM. http://www
.iitp.org.nz/files/wipsce-teachers-2013.pdf.

Thompson, David, Tim Bell, Peter Andreae, and Anthony Robins. 2013. "The Role of
Teachers in Implementing Curriculum Changes." In *SIGCSE 2013—Proceedings of the
44th ACM Technical Symposium on Computer Science Education*, edited by Tracy Camp,
Paul Tymann, J. D. Dougherty, and Kris Nagel, 245–250. New York: ACM.

University of Auckland. 2021. "Stem Online NZ." https://www.stemonline.auckland
.ac.nz/.

Victorian Curriculum and Assessment Authority. 2021. "VCE Study Designs."
https://vcaa.vic.edu.au/curriculum/vce/vce-study-designs/Pages/vce-study-designs
.aspx.

6

COMPUTATIONAL THINKING ASSESSMENT: A DEVELOPMENTAL APPROACH

Marcos Román-González and Juan-Carlos Pérez-González

INTRODUCTION

In a previous work (Román-González, Moreno-León, and Robles 2019), we argued for the need to build inclusive systems of assessments in order to perform comprehensive evaluations of educational interventions that involve computational thinking (CT). To do so, we started by proposing a taxonomy of CT assessment tools regarding their evaluative approach. Thus, we could differentiate between seven types of instruments: (1) *diagnostic tools* (i.e., those that measure the CT aptitudinal level of the subject); (2) *summative tools* (i.e., those that evaluate if the learner has achieved enough content knowledge—and/or if he is able to perform properly—after receiving some instruction or training in CT skills); (3) *formative-iterative tools* (i.e., those that provide feedback to the learner to improve his/her CT skills); (4) *data-mining tools* (i.e., those that retrieve and record the learner activity in real time while acquiring computational concepts and practices); (5) *skill transfer tools* (i.e., those that assess to what extent the students are able to transfer their CT skills onto different kinds of problems, contexts, and situations); (6) *perceptions–attitudes scales* (i.e., those that are aimed at assessing the perceptions, such as self-efficacy perceptions, and attitudes of the subjects not only about CT, but also about related issues such as computers, computer science, computer programming, or even digital literacy); and (7) *vocabulary assessment tools* (i.e., those that intend to measure several elements and dimensions of CT when they are verbally expressed by the subjects).

Moreover, in that previous work (Román-González, Moreno-León, and Robles 2019), we established three possible criteria that could provide guidance on how to properly combine the aforementioned types of CT assessment instruments. Hence, we first explored to what extent each type of tool is suitable for measuring distinct CT dimensions (i.e., computational concepts, computational practices, and computational perspectives). Second, we exposed how to use each of those kinds of instruments regarding the different phases and chronological points within a CT educational evaluation (i.e., before, along, just after, or sometime after the intervention). Finally, we attempted to state what level(s) of Bloom's (revised) taxonomy of cognitive processes each type of CT assessment tools is addressing. However, nothing was said about how to align this myriad of CT assessment tools with the various ages and developmental stages of the individuals along K–12 education. Indeed, there is a notorious theoretical and empirical gap regarding CT assessment from a developmental perspective, and only occasional research has been undertaken in this vein.

In sum, the main contribution of that previous chapter (Román-González, Moreno-León, and Robles 2019) was to define an inclusive framework that enlightens how to combine different CT assessment tools to conduct comprehensive evaluations *at a certain moment*. In other words, our previous work was written from a *cross-sectional perspective* since its goal was to show how to set up comprehensive research designs *at a certain moment*. Nevertheless, we did not address how to articulate CT assessment from a developmental approach; that is, how to conduct CT assessments along the successive stages of human development. This is precisely the aim of the present chapter. To achieve our goal, we will intersect some of the current corpus of scientific knowledge on CT assessment with Piagetian and neo-Piagetian developmental theories, along four school stages (kindergarten, elementary school, middle school, and high school). For each of the four stages, we will try to answer the following questions:

- Which CT cognitive subprocesses are mainly involved and may be assessed at each stage of development? Which other basic cognitive skills may support CT at each stage of development?
- Consequently, which computational concepts, practices, and perspectives may be specifically addressed at each stage of development?
- Finally, which kind of CT assessment instruments may be more appropriate at each stage of development? Can we find good examples of CT assessment tools at each stage?

Ultimately, when answering the prior questions, the goal of the present chapter is to sketch a possible framework that could guide the configuration of CT longitudinal research designs. Therefore, the present manuscript is written from a *longitudinal perspective*.

BACKGROUND: CT FROM A COGNITIVE PERSPECTIVE

In previous papers, we have offered several definitions of computational thinking (CT). For instance, we have stated that CT involves the ability to formulate and solve problems by relying on the fundamental concepts and practices of computing (Román-González, Pérez-González, and Jiménez-Fernández 2017). We have also defined CT as the ability that allows the subject to effectively solve problems and express ideas by using the power of computers (Moreno-León, Robles, Román-González, and Rodríguez García 2019). However, none of the aforementioned definitions is enough to face the target of the present chapter. Since we aim to intersect CT with Piagetian developmental theories, it is necessary to define CT in cognitive terms.

In this vein, CT can be understood as a problem-solving ability that is composed by the following series of successive cognitive subprocesses (figure 6.1), namely:

i. **Decomposition**: to break down a problem into smaller and simpler elements.

ii. **Pattern recognition**: to perceive and detect regularities between those elements.

iii. **Abstraction**: to remove/ignore nonrelevant details and information of the problem to highlight the critical variables that will enable one to represent it properly (i.e., often called "internal generalization").

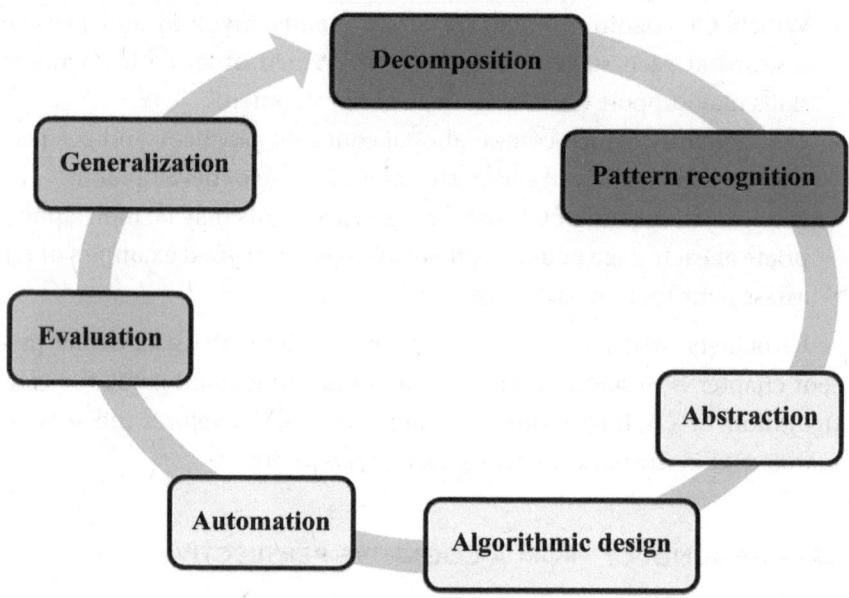

6.1 Cognitive subprocesses involved in CT.

iv. **Algorithmic design**: to build step-by-step instructions that, when followed, will allow for solving the problem.
v. **Automation**: to implement the aforementioned instructions within a digital device by means of some kind of computer programming.
vi. **Evaluation**: to assess the efficacy and efficiency of that implementation in order to debug and/or to improve the programming code.
vii. **Generalization**: to transfer the achieved solution to a wider range of analogous problems (i.e., often called "external generalization").

We can exemplify the previous figure trying to think computationally on the following problem: "what clothes should I wear today?" If we want to project CT on that problem, firstly we must *decompose* the broad category "clothes" into simpler elements (e.g., underwear, bottom clothes, top clothes, shoes, and accessories). Then, we must recognize some *patterns* along those elements (e.g., long trousers are usually worn with shirts, while short trousers do so with t-shirts; umbrellas are only used while raining, while coats only do so under certain temperature; blue and white clothes tend to fit together, while red and green do not). Afterwards, it becomes essential to detect and *abstract* the critical variables that permit to internally represent the space-problem (e.g., the clothes I should wear today will probably depend on "the day of the week" (working day versus weekend), "the weather" (e.g., temperature or presence or absence of rainfalls), "the range of colors" I wish to wear, and so on. Then, it becomes possible to *design an algorithm* based on those critical variables, which could state the instructions to answer the problem in different conditions. For example:

If today is a working day, I should wear long trousers, a shirt, and a jacket; else [if temperature is over 20°C, I should wear short trousers and a t-shirt; else I should wear a tracksuit].

Once the algorithm has been designed, it can be *automated* and implemented in a digital device through programming and executing some computer code (e.g., we can implement the aforementioned algorithm that solves "what clothes should I wear today" as a mobile app written in App Inventor language, which could be connected to the calendar and to the weather forecast to give an answer to that question). Furthermore, the algorithm and its corresponding programming code can be *evaluated* and refined to improve their efficacy and efficiency (e.g., the algorithm may better fix the values of some parameters and/or may include new variables such as "the current mood" of the subject who is using the clothing app). Finally, we can perform a second-level abstraction to find some communalities between our specific problem and a wider family of analogous ones, so it will become possible to transfer and externally *generalize* some elements of our specific solution (e.g., we may generalize our algorithm and its corresponding programming code to a similar problem such as "what food should I eat today?").

It is worth noting that not all the aforementioned cognitive subprocesses stand at the same level of importance and hierarchy within the whole CT process. On the one hand, *decomposition* and *pattern recognition* (see the green boxes in figure 6.1) are common elements in almost every problem-solving task. In other words, "decomposition" and "pattern recognition" are not specific to CT, and they could even be considered just as prerequisites to think computationally.

On the other hand, *abstraction, algorithmic design,* and *automation* (see the blue boxes in figure 6.1) can be located at the core of the CT process. Thus, relevant authors have stated that abstraction (e.g., Grover and Pea

2013; Wing 2006) and/or algorithmic thinking (Aho 2012; Shute, Sun, and Asbell-Clarke 2017) are the main cognitive subprocesses for thinking computationally. For example, Jeanette Wing affirms that CT "requires thinking at multiple levels of abstraction" (Wing 2006, 35), and/or Alfred Aho defines CT as "the thought process involved in formulating problems so their solutions can be represented as computational steps and algorithms" (Aho 2012, 832). Nevertheless, if we only take into account "abstraction" and "algorithmic design" to define the core of CT, then it could become indistinguishable from mathematical thinking (Stephens and Kadijevich 2020). According to these authors and from our point of view, it is indispensable to add "automation" as a third element to the aforementioned pair in order to clearly characterize the essence of CT.

Finally, *evaluation* and *generalization* (see the red boxes in figure 6.1) may be considered rather as consequences, implications, or applications of CT than as central elements of it.

Furthermore, if we focus again on *algorithmic design* as the most specific cognitive subprocess of CT, then it comes very relevant to point out what structures can be progressively learned and used by the individual to build better and more complex algorithms. These structures are (Mühling, Ruf, and Hubwieser 2015; Román-González 2016): (a) *sequencing* structures; (b) *repetition* structures; (c) *selection* structures; (d) *modularization* structures; and (e) *parameterization* structures (figure 6.2).

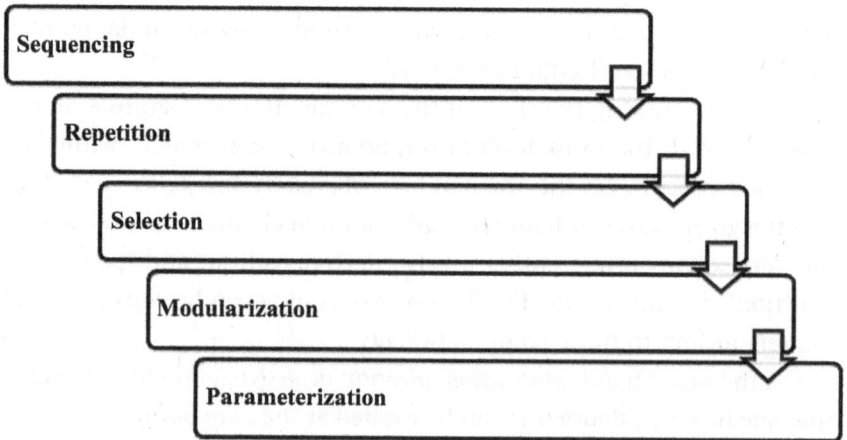

6.2 Algorithmic design structures.

The aforementioned structures for algorithmic design involve increasing levels of abstraction, they can be progressively incorporated and nested along algorithmic solutions, and they can all be automated by means of computer programming.

In the following sections, we will take figures 6.1 and 6.2 as a reference to describe CT and its assessment from a developmental approach.

CT ASSESSMENT FROM A DEVELOPMENTAL APPROACH

Once we have defined CT in cognitive terms, it becomes possible to intersect CT with Piaget's theory of cognitive development. The goal of this intersection is to explore and describe CT assessment from a developmental approach, along the following four educational stages: kindergarten, elementary school, middle school, and high school.

CT ASSESSMENT IN KINDERGARTEN

Kindergarten mainly concurs with the so-called preoperational stage in Piagetian theory (Ginsburg and Opper 1988). During this stage, young children are not able to perform logical operations, nor onto mental concepts nor even onto physical elements that surround the subjects. Consequently, kindergarteners cannot abstract or think in algorithmic terms, so these core elements of CT should not be assessed during this stage.

In other words, during kindergarten, we should not expect to develop and assess the whole cycle of CT cognitive subprocesses depicted in figure 6.1. At most, we may focus on fostering and assessing so-called prerequisites of CT, namely "decomposition" and "pattern recognition." In this vein, some basic cognitive abilities that become critical to support CT development at this stage are attentional and perceptual skills (Georgiou and Angeli 2019; Marinus et al. 2018; Urlings, Coppens, and Borghans 2019) (see figure 6.11).

Referring to figure 6.2, on the one hand, just some sequencing and repetition protostructures may be found during this stage. These protostructures rely on and are limited by one main characteristic of preoperative thought: irreversibility (i.e., when children are unable to mentally reverse a sequence of events). Thus, kindergarteners are only able to break down, serialize, and sequence elements in one single direction. On the other hand, selection structures are unexpected to appear during kindergarten, since precausal thinking is another main feature of children in the preoperational stage.

Finally, we can state some principles (and cite some good examples) for addressing CT assessment in kindergarten:

- Although kindergarteners are not able to perform logical operations, they can represent and understand symbols, mainly by means of so-called symbolic play. In this vein, CT assessment should rely on playing symbolic games that involve CT prerequisites and sequencing/repetition protostructures (e.g., when assessing CT within the context of playing with Bee-Bots or KIBO robots, which in addition align with the "animism" of kindergarteners) (Critten, Hagon, and Messer 2021; Kotsopoulos et al. 2021; Relkin, de Ruiter, and Bers 2021).
- Since these children are not able to abstract information, paper-and-pencil assessment or testing are not recommended along this stage. Instead of that, it may be more appropriate to assess CT through observation grids and templates while kids are directly manipulating physical objects to solve problems that involve CT prerequisites (Angeli and Valanides 2020; Diago, Arnau, and González-Calero 2018).
- Since kindergarteners are keen on learning by means of narratives and tales, one relevant and natural way to develop and to assess CT along this stage is through decomposing and sequencing stories (Kazakoff, Sullivan, and Bers 2013; Terroba, Ribera, and Lapresa 2020).

CT ASSESSMENT IN ELEMENTARY SCHOOL

Elementary school mainly concurs with the so-called concrete operational stage in Piagetian theory (Ginsburg and Opper 1988). During this stage, children are able to solve problems and perform logical operations onto concrete and specific objects/events that are within their reach. In other words, these individuals are capable of inductive reasoning based on concrete and specific elements around them, often by means of trial-and-error strategies. Conversely, elementary school students have not yet acquired or consolidated deductive reasoning, which involves using general principles to hypothesize and predict further results. Retrieving the terms already used in our background section, we may say that elementary school students can perform first-level abstractions (often called "internal or inductive generalizations"), but not second-level abstractions (often called "external or deductive generalizations").

Referring to figure 6.1, during elementary school, children start to deal with the core elements of CT (namely "abstraction," "algorithmic design," and "automation"). Since elementary school students are able to perform first-level abstractions, they are ready to start designing and automating simple algorithms, which will become more complex along the stage. Then, those core elements of CT should be developed and assessed during elementary school. Conversely, since these children are not able to perform second-level abstractions, the CT cognitive subprocess named "generalization" can hardly be developed or assessed within this stage. Furthermore, since elementary schoolers have not yet acquired enough metacognitive skills, it also does not seem appropriate to assess the "evaluation" subprocess.

Referring to figure 6.2, during elementary school, children can learn and properly use sequencing and repetitions structures (corresponding with computational concepts such as "repeat times-loop" or "repeat until-loop"). Moreover, since children at this age have already acquired causal thinking, they can also understand and apply selection structures (corresponding with computational concepts such as "if-conditional" or "if/else-conditional"). In contrast, it is not to be expected that elementary schoolers use modularization and parameterization structures properly due to the high degree of formalization of these structures, which correspond with computational concepts such as "functions" and "variables."

Nevertheless, in some other works, we have qualified what is said in the previous paragraph:

• Due to the maturational limitations of working memory, elementary school students may struggle when mentally sequencing long series of objects/events (a number of four to five elements seems to be the upper limit). These children may also have difficulties when using the "while-loop" since it requires them to apply a repetition structure while a certain condition is met, which is consequently a very demanding computational concept for the working memory of elementary schoolers (Zapata-Cáceres, Martín-Barroso, and Román-González 2021).

- There is evidence that numerical factor/ability is critical to support the development of repetition structures during elementary school (Tsarava et al. 2022) (see figure 6.11).
- There is also evidence that visual elements and colors can scaffold the acquisition of difficult repetition and selection structures in elementary school students.

An excellent example of a CT assessment tool aimed at elementary school is the *Beginners Computational Thinking Test (BCTt)* (Zapata-Cáceres, Martín-Barroso, and Román-González 2020). This test consists of twenty-five items and is aligned with all the guidelines exposed along this subsection (figures 6.3 and 6.4).

CT ASSESSMENT IN MIDDLE SCHOOL

Middle school coincides with the beginning of the "formal operational stage" in Piaget's theory of cognitive development (Ginsburg and Opper 1988). During middle school, individuals start to perform logical operations onto symbols related to abstract concepts. Consequently, hypothetical

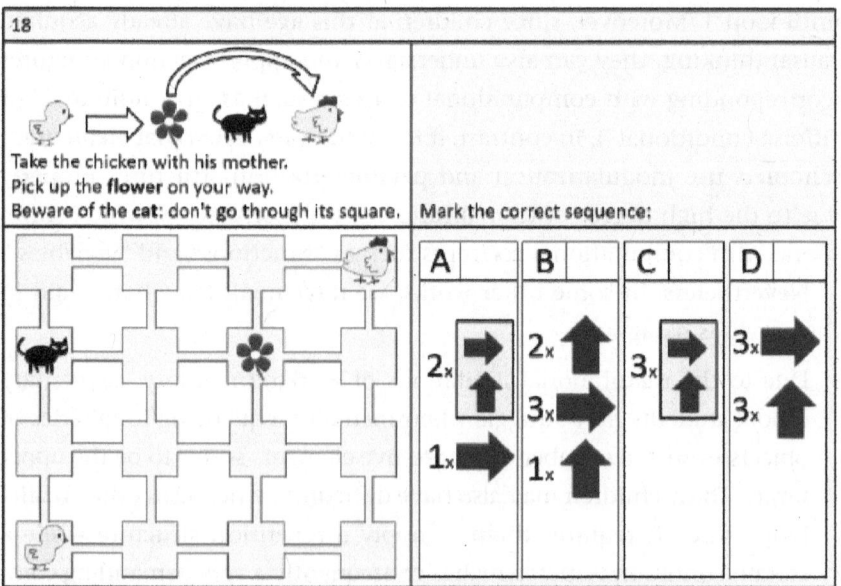

6.3 *BCTt* item example (item #18).

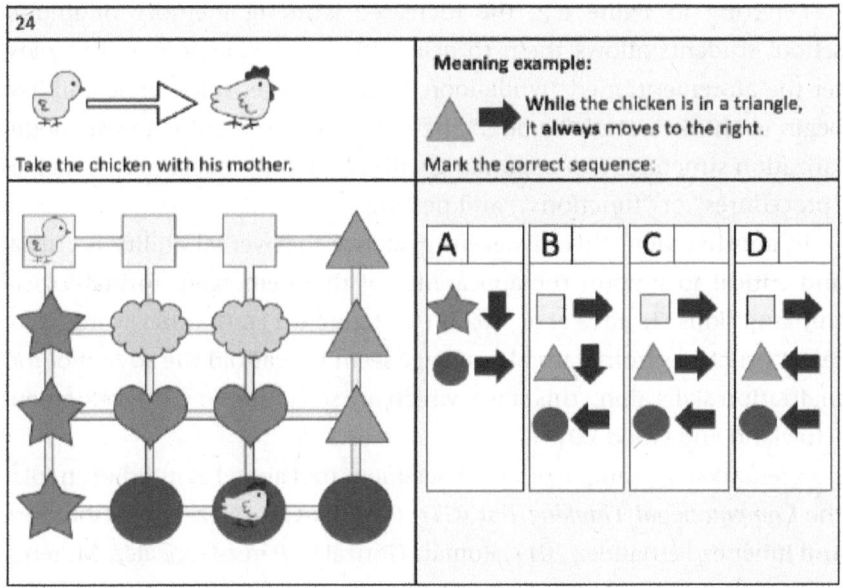

6.4 *BCTt* item example (item #24).

and deductive reasoning emerges during middle school. Moreover, three cognitive milestones can be highlighted in middle school students:

- The emergence of abstract thinking occurs, which goes beyond concrete and specific objects and events.
- Middle schoolers develop metacognitive skills, which allow the subject to consciously observe, to reason about, and to supervise his/her own thinking.
- Problem-solving in middle schoolers becomes systematical, since they begin to solve problems in a logical and methodical way (not just by means of trial-and-error strategies).

Overall, middle school students begin to distance themselves from concrete reality and from their own cognitive processes, which is also supported by an increase of their working memory and processing capacity. All of the above have obvious consequences on CT development along this stage. Referring to figure 6.1, middle school students become capable of evaluating their own algorithmic solutions (and their corresponding programming codes). Thus, "evaluation" is a CT cognitive subprocess that should be assessed from this stage onwards, mainly by means of debugging computational practices.

Referring to figure 6.2, the increased working memory of middle school students allows them to deal with long sequences and to master the aforementioned "while-loop." Furthermore, since these children begin to think "out of the box," they become also capable to use modularization structures (corresponding with computational concepts such as "procedures" or "functions") and nesting.

In another vein, there is recent evidence that verbal ability is crucial and critical to support the appearance of these emergent formal-logical thinking skills (Tsarava et al. 2022) (see figure 6.11). In other words, syntactic structures from natural language seem to scaffold the advent of formalization skills along this stage, which are so central to CT development (Howland and Good 2015).

A remarkable example of a CT assessment tool aimed at middle school is the *Computational Thinking Test (CTt)* (Román-González, Pérez-González, and Jiménez-Fernández 2017; Román-González, Pérez-González, Moreno-León, and Robles 2018a). This test consists of twenty-eight items and is aligned with all the ideas exposed along this subsection (figures 6.5 and 6.6).

CT ASSESSMENT IN HIGH SCHOOL (AND BEYOND)

High school often involves the consolidation of the "formal operational stage" in Piaget's theory (Ginsburg and Opper 1988). During this stage, individuals sharpen their formal-logical abilities onto abstract concepts,

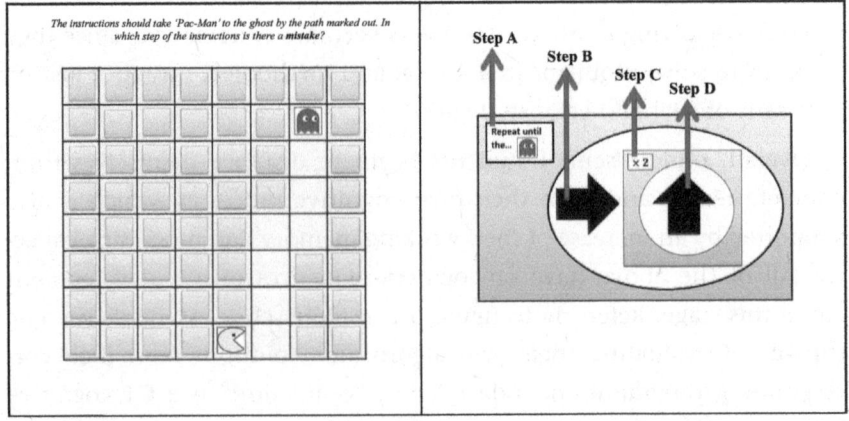

6.5　*CTt* item example (item #11).

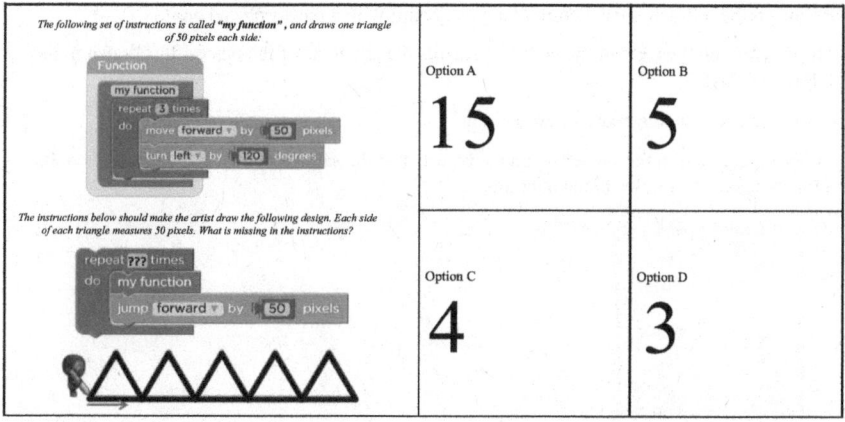

The following set of instructions is called **"my function"** , and draws one triangle of 50 pixels each side:

	Option A	Option B
	15	5

The instructions below should make the artist draw the following design. Each side of each triangle measures 50 pixels. What is missing in the instructions?

	Option C	Option D
	4	3

6.6 *CTt* item example (item #26).

and deductive-hypothetical reasoning skills are refined. Retrieving our previous terms, we can affirm that high school students are finally capable of performing second-order abstractions (i.e., to find some communalities between a specific problem and a wider family of analogous ones), so they become able to transfer and externally generalize some elements of their algorithmic solutions. In this vein and referring to figure 6.1, the CT cognitive subprocess called "generalization" should be specially developed and assessed along this stage.

Referring to figure 6.2, refined formal-logical abilities of high school students permit them to understand and use parameterization structures (corresponding with computational concepts such as "functions with parameters" and "variables"[1]). In another vein, recent evidence suggests that nonverbal reasoning (also called visual or figurative reasoning) is critical to foster and consolidate CT along this final stage (Tsarava et al. 2022) (see figure 6.11).

When we search for good examples of CT assessment tools aimed at high school (and beyond), the results are scarce. In this regard, one work in progress is the *Computational Thinking Test for Higher Education*[2] *(CTt-H)* (Lafuente Martínez et al. 2022), which intends to assess the transfer of CT on a wide variety of problems (figures 6.7 and 6.8).

Finally, it is also worth noting that neo-Piagetian theory declares the existence of one more stage in human development called the "postformal thought stage" (e.g., Sinnott 1998). Postformal thought involves the

Below on the left you see a picture of a game board with 4 pieces placed on it.

The diagram on the right of the board represents the positions of the pieces. It is drawn in the following way:

- For each piece on the board, draw a circle.

- If two pieces are in the same row on the board or in the same column on the board, then draw a line between the circles in the diagram.

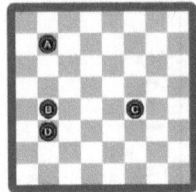

Letters have been placed and the circles so you easily check that the diagram is correct.

A new board of six pieces is shown below.

A new position diagram for this board is drawn in the same way.

Question:
Which of the four diagrams below were drawn?

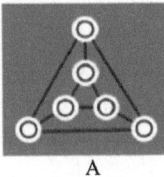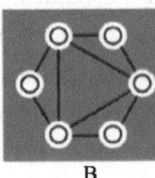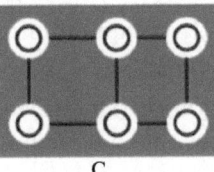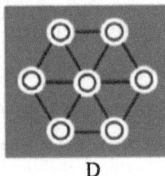

A   B   C   D

6.7   *CTt-H* item example #1.

Five people numbered 1, 2, 3, 4, 5 are trying to cross a road that contains a number of deep holes. All holes have in depth 3. To pass over holes with depth 3, the 3 leading people have to go into the holes in order so that the other people can safely cross the hole. Then, the last people who crossed the hole will pull the highest people from the hole up, and so on. For clarity, look at the diagram below where these people are crossing the first hole:

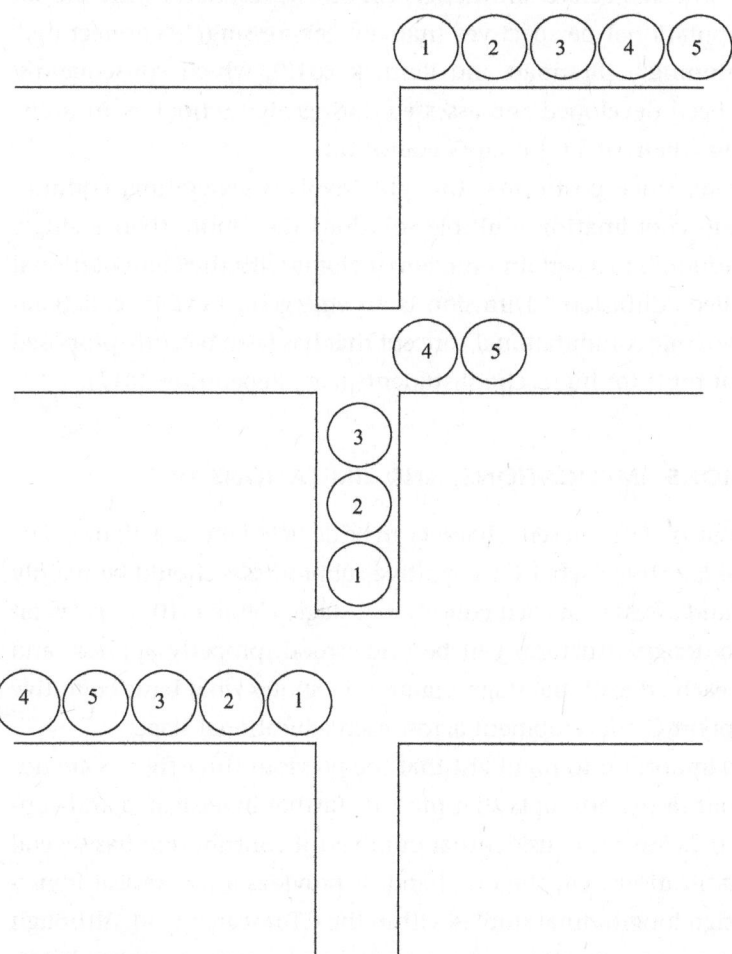

After which hole will they reach the intial formation again (1, 2, 3, 4, 5)?

A) 11th hole
B) 15th hole
C) 16th hole
D) 25th hole

6.8 *CTt-H* item example #2.

communication and coordination of multiple different logics in a dialec-
tical and flexible way. Thus, postformal thought allows the individuals
to coordinate multiple goals, methods, causalities, and results to reach a
deeper and intersubjective knowledge. All of this resonates with the so-
called computational perspectives (namely "expressing," "connecting,"
and "questioning") (Brennan and Resnick 2012), which consequently
must have been developed and assessed during high school as an indispensable ingredient of a CT quality education.

Even more, since postformal thought involves generating, communicating, and coordinating multiple solutions (i.e., more than a single
"correct solution") to a certain problem, it aligns with the computational
concept called "diffusion." Diffusion is an emergent, flexible, collaborative, and dynamic computational concept that has been recently proposed
as a relevant topic for high school students (e.g., Repenning 2017).

CONCLUSIONS, IMPLICATIONS, AND LIMITATIONS

The conclusions of the present chapter can be depicted through figures 6.9–
6.11. Figure 6.9 shows what CT cognitive subprocesses should be mainly
developed and assessed at each educational stage. Figure 6.10 shows what
algorithmic design structures can be understood, properly applied, and
assessed at each educational stage. Figure 6.11 shows what basic cognitive
abilities support CT development across each educational stage.

It is very important to highlight that the previous three figures are not
definitive but rather attempts that must be further investigated and contrasted. In any case, we consider that our present contribution has several
relevant implications. On the one hand, it provides a theoretical framework to design longitudinal studies within the CT research field. Although
they are more expensive due to experimental and sample mortality, longitudinal studies usually lead to more valid results than cross-sectional studies, since the former take into account the same subjects along time while
the latter simultaneously utilize different cohorts of individuals (which
implies some threats to the validity of their results). On the other hand, if
a CT developmental theory is empirically confirmed, that will contribute
to reinforce the construct validity of CT. There is still a long way to go.

Finally, we must express some limitations to our proposal that coin-
cide with the limitations that have been pointed out to Piagetian theory

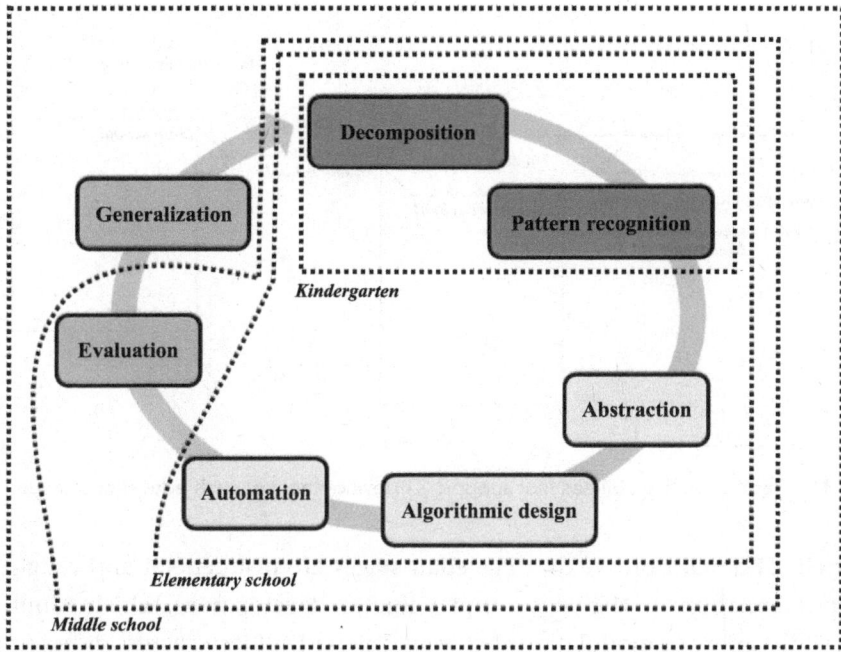

6.9 CT cognitive subprocesses at each educational stage.

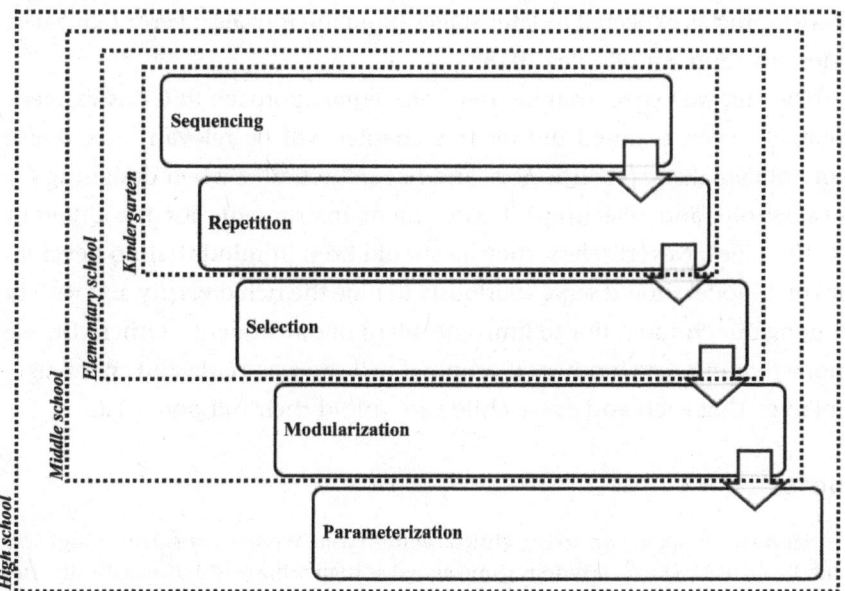

6.10 Algorithmic design structures at each educational stage.

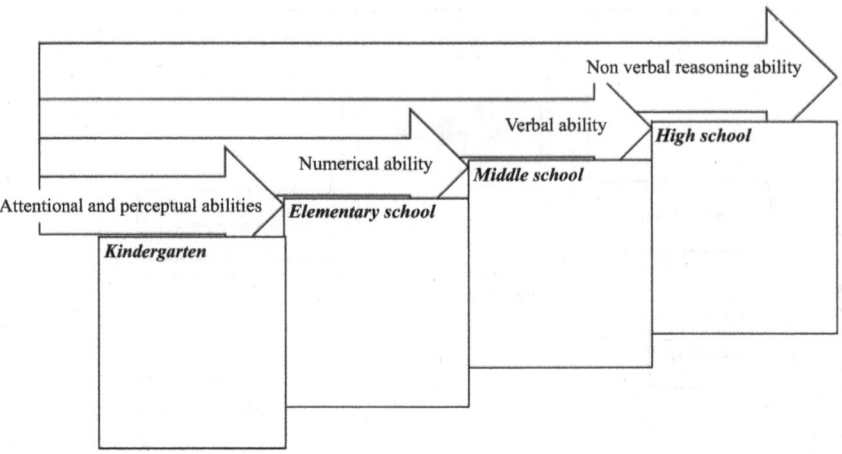

6.11 Basic cognitive abilities that support CT development at each educational stage.

itself. The main one is that Piagetian stages are just general and rough approximations to the very complex human development, which admit a wide variety of modulations between different subjects, across different cultures, and across different domains of knowledge and expertise. In this vein and in relation to the CT domain, we have found and studied cases of Spanish "computational talents" in middle school who are capable of performing as expected in later stages (Román-González, Pérez-González, Moreno-León, and Robles 2018b).

Overall, we expect that the developmental approach to CT assessment that has been exposed during this chapter will be relevant and useful for policymakers, practitioners, and researchers alike when designing CT evaluations and selecting CT assessment instruments for the different K–12 grades. Nevertheless, they all should keep in mind that no developmental model should serve spuriously to hide the rich diversity and variety among our children nor to limit the talent of our students. Otherwise, we hope that our developmental proposal will serve as guide and encouragement so that each and every child can unfold their full potential.

NOTES

1. Recently in Spain, *Articoding* (https://github.com/WeArePawns/Articoding) was developed. *Articoding* is a serious game aimed at high schoolers that specifically uses and relies on "variables" as an anchor to teach all the rest of computational concepts (Faouaz, García, and Poyatos 2021).

2. The initial pool of items under validation is available at: https://www.surveymonkey.com/r/DN9V7YW.

REFERENCES

Aho, Alfred V. 2012. "Computation and Computational Thinking." *The Computer Journal* 55, no. 7: 832–835. https://doi.org/10.1093/comjnl/bxs074.

Angeli, Charoula, and Nicos Valanides. 2020. "Developing Young Children's Computational Thinking with Educational Robotics: An Interaction Effect Between Gender and Scaffolding Strategy." *Computers in Human Behavior* 105: 105954. https://doi.org/10.1016/j.chb.2019.03.018.

Brennan, Karen, and Mitchel Resnick. 2012. "New Frameworks for Studying and Assessing the Development of Computational Thinking." In *Proceedings of the 2012 Annual Meeting of the American Educational Research Association (AERA '2012)*, 1–25. http://scratched.gse.harvard.edu/ct/files/AERA2012.pdf.

Critten, Valerie, Hannah Hagon, and David Messer. 2021. "Can Pre-school Children Learn Programming and Coding Through Guided Play Activities? A Case Study in Computational Thinking." *Early Childhood Education Journal*. https://doi.org/10.1007/s10643-021-01236-8.

Diago, Pascual D., David Arnau, and José Antonio González-Calero. 2018. "La resolución de problemas matemáticos en primeras edades escolares con Bee-Bot." *Matemáticas, Educación y Sociedad* 1, no. 2: 36–50. https://www.uco.es/ucopress/ojs/index.php/mes/article/view/12835.

Faouaz, Dany, Arturo García, and Álvaro Poyatos. 2021. *Serious Games to Promote Computational Thinking and Coding* [Bachelor Thesis]. Madrid: UCM. https://eprints.ucm.es/id/eprint/66940/.

Georgiou, Kyriakoula, and Charoula Angeli. 2019. "Developing Preschool Children's Computational Thinking with Educational Robotics: The Role of Cognitive Differences and Scaffolding." In *Proceedings of the 2019 International Conference on Cognition and Exploratory Learning in the Digital Age (CELDA '2019)*, 101–108. http://dx.doi.org/10.33965/celda2019_201911L013.

Ginsburg, Herbert P., and Sylvia Opper. 1988. *Piaget's Theory of Intellectual Development (3rd Ed.)*. New Jersey: Prentice-Hall.

Grover, Shuchi, and Roy Pea. 2013. "Computational Thinking in K–12: A Review of the State of the Field." *Educational Researcher* 42, no. 1: 38–43. https://doi.org/10.3102/0013189X12463051.

Howland, Kate, and Judith Good. 2015. "Learning to Communicate Computationally with Flip: A Bi-Modal Programming Language for Game Creation." *Computers & Education* 80: 224–240. https://doi.org/10.1016/j.compedu.2014.08.014.

Kazakoff, Elizabeth R., Amanda Sullivan, and Marina U. Bers. 2013. "The Effect of a Classroom-Based Intensive Robotics and Programming Workshop on Sequencing

Ability in Early Childhood." *Early Childhood Education Journal* 41: 245–255. https://doi.org/10.1007/s10643-012-0554-5.

Kotsopoulos, Donna, Lisa Floyd, Brandon A. Dickson, Vivian Nelson, and Samantha Makosz. 2021. "Noticing and Naming Computational Thinking during Play." *Early Childhood Education Journal.* https://doi.org/10.1007/s10643-021-01188-z.

Lafuente Martínez, , M., Olivier Lévêque, Isabel Benítez, Cécile Hardebolle, and Jessica Dehler Zufferey. 2022. "Assessing Computational Thinking: Development and Validation of the Algorithmic Thinking Test for Adults." *Journal of Educational Computing Research* 60, no. 6: 1436–1463. https://doi.org/10.1177/07356331211057819.

Marinus, Eva, Zoe Powell, Rosalind Thornton, Genevieve McArthur, and Stephen Crain. 2018. "Unravelling the Cognition of Coding in 3-to-6-Year Olds: The Development of an Assessment Tool and the Relation between Coding Ability and Cognitive Compiling of Syntax in Natural Language." In *Proceedings of the 2018 ACM Conference on International Computing Education Research (ICER '18),* 133–141. https://doi.org/10.1145/3230977.3230984.

Moreno-León, Jesús, Gregorio Robles, Marcos Román-González, and Juan David Rodríguez García. 2019. "Not the Same: A Text Network Analysis on Computational Thinking Definitions to Study Its Relationship with Computer Programming." *Revista Interuniversitaria de Investigación en Tecnología Educativa* 7: 26–35. https://doi.org/10.6018/riite.397151.

Mühling, Andreas, Alexander Ruf, and Peter Hubwieser. 2015. "Design and First Results of a Psychometric Test for Measuring Basic Programming Abilities." In *Proceedings of the Workshop in Primary and Secondary Computing Education (WiPSCE '15),* 2–10. https://doi.org/10.1145/2818314.2818320.

Relkin, Emily, Laura E. de Ruiter, and Marina U. Bers. 2021. "Learning to Code and The Acquisition of Computational Thinking by Young Children." *Computers & Education* 169: 104222. https://doi.org/10.1016/j.compedu.2021.104222

Repenning, Alexander. 2017. "Moving Beyond Syntax: Lessons from 20 Years of Blocks Programing in AgentSheets." *Journal of Visual Language and Sentient Systems* 3: 68–91. https://dblp.org/rec/journals/jvlc/Repenning17.

Román-González, Marcos. 2016. *Codigoalfabetización y pensamiento computacional en Educación Primaria y Secundaria: validación de un instrumento y evaluación de programas* [doctoral thesis]. Madrid: UNED. http://e-spacio.uned.es/fez/view/tesisuned:Educacion-Mroman.

Román-González, Marcos, Jesús Moreno-León, and Gregorio Robles. 2019. "Combining Assessment Tools for a Comprehensive Evaluation of Computational Thinking Interventions." In *Computational Thinking Education,* edited by Siu-Cheung Kong and Harold Abelson, 79–98. Singapore: Springer. https://doi.org/10.1007/978-981-13-6528-7_6.

Román-González, Marcos, Juan-Carlos Pérez-González, and Carmen Jiménez-Fernández. 2017. "Which Cognitive Abilities Underlie Computational Thinking?

Criterion Validity of the Computational Thinking Test." *Computers in Human Behavior* 72: 678–691. https://doi.org/10.1016/j.chb.2016.08.047.

Román-González, Marcos, Juan-Carlos Pérez-González, Jesús Moreno-León, and Gregorio Robles. 2018a. "Extending the Nomological Network of Computational Thinking with Non-Cognitive Factors." *Computers in Human Behavior* 80: 441–459. https://doi.org/10.1016/j.chb.2017.09.030.

Román-González, Marcos, Juan-Carlos Pérez-González, Jesús Moreno-León, and Gregorio Robles. 2018b. "Can Computational Talent Be Detected? Predictive Validity of the Computational Thinking Test." *International Journal of Child-Computer Interaction* 18: 47–58. https://doi.org/10.1016/j.ijcci.2018.06.004.

Shute, Valerie J., Chen Sun, and Jodi Asbell-Clarke. 2017. "Demystifying Computational Thinking." *Educational Research Review* 22: 142–158. https://doi.org/10.1016/j.edurev.2017.09.003.

Sinnott, Jan D. 1998. *The Development of Logic in Adulthood. Postformal Thought and Its Applications*. Boston, MA: Springer. https://doi.org/10.1007/978-1-4757-2911-5.

Stephens, Max, and Djordje M. Kadijevich. 2020. "Computational/Algorithmic Thinking." In *Encyclopedia of Mathematics Education*, edited by Stephen Lerman, 117–123. Cham: Springer. https://doi.org/10.1007/978-3-030-15789-0_100044.

Terroba, Marta, Juan Miguel Ribera, and Daniel Lapresa. 2020. "Pensamiento computacional en la resolución de problemas contextualizados en un cuento en Educación Infantil." *Edma 0–6: Educación Matemática en la Infancia* 9, no. 2: 73–92. http://www.edma0-6.es/index.php/edma0-6/article/view/156.

Tsarava, Katerina, Korbinian Moeller, Marcos Román-González, Jessika Golle, Luzia Leifheit, Martin V. Butz, and Manuel Ninaus. 2022. "A Cognitive Definition of Computational Thinking in Primary Education." *Computers & Education* 179: 104425. https://doi.org/10.1016/j.compedu.2021.104425.

Urlings, Corrie C., Karien M. Coppens, and Lex Borghans. 2019. "Measurement of Executive Functioning Using a Playful Robot in Kindergarten." *Computers in the Schools* 36, no. 4: 255–273. https://doi.org/10.1080/07380569.2019.1677436.

Wing, Jeannette M. 2006. "Computational Thinking." *Communications of the ACM* 49, no. 3: 33–35. https://doi.org/10.1145/1118178.1118215.

Zapata-Cáceres, María, Estefanía Martín-Barroso, and Marcos Román-González. 2020. "Computational Thinking Test for Beginners: Design and Content Validation." In *Proceedings of the 2020 IEEE Global Engineering Education Conference (EDUCON '2020)*, 1905–1914. https://doi.org/10.1109/EDUCON45650.2020.9125368.

Zapata-Cáceres, María, Estefanía Martín-Barroso, and Marcos Román-González. 2021. "Computational Thinking Collaborative Game-Based Learning Environment and Assessment Tool: A Case Study in Primary School." *IEEE Transactions on Learning Technologies* 14, no. 5: 576–589. https://doi.org/10.1109/TLT.2021.3111108.

7

PATHWAYS OF COMPUTING EDUCATION: FORMAL AND INFORMAL APPROACHES

Chee-Kit Looi, Shiau-Wei Chan, Peter Seow, Longkai Wu, and Bimlesh Wadhwa

INTRODUCTION

Today, computing is becoming gradually more essential to our society. In many countries, it has been introduced into compulsory schooling (K–12) education, including in the form of STEM (Science, Technology, Engineering, and Mathematics) education. Manches and Plowman (2017) argue that computing education should be introduced to children from an early age. Computing is often taught in schools in the form of computational thinking (CT), which is about expressing solutions as algorithms or computational steps that can be executed using a computer (CSTA 2016). CT is advocated as a universal competence that can prepare children for future challenges in a growing digital world (Voogt et al. 2015). It is defined by Wing (2006) as "solving problems, designing systems, and understanding human behaviour, by drawing on the concepts fundamental to computer science" (33). In computing education, the definition of CT has two aims—(1) learning transferable knowledge from computing that can be utilized in everyday life and (2) employing computing concepts to promote computing work in other subjects (Guzdial 2015). The ideas of programming and algorithms that are mostly used to assess CT in K–12 computing education include variables, modularity, control, and algorithms (Alves, Von Wangenheim, and Hauck 2019).

Different learning pathways could meet students' needs in more effective ways than a single learning channel. Computationally talented learners show different abilities when learning coding or programming compared to regular learners; for instance, they can accelerate from block-based programming to text-based programming (Roman-Gonzalez et al. 2018). Learners with disabilities in computing require additional CT-specific supports to engage in the CT instruction, such as daily behavior reports, voluntary breaks, and after-school make-up time (Snodgrass, Israel, and Reese 2016).

From a macro perspective, formal and informal learning systems can be regarded as the two main learning pathways in computing education. Generally, a formal learning system can deliver a structure for systematic thinking and methods, while an informal learning system can assist students to increase their motivation and to recognize their interests (Bocconi et al. 2016). Many countries have infused computer science (CS) into K–12 formal education, including England, Estonia, Finland, Poland, and Slovakia. In addition to these formal learning environments, many informal offerings aim to increase children's knowledge and interest of CS. Even if they are not part of the curriculum, they also provide opportunities to communicate with CS as they mainly focus on coding or programming, for instance, the website Code.org and the programming clubs of Coder Dojo and Code Club. The learners can participate in these extracurricular activities that are not available in a regular classroom environment (Lunenburg 2010).

Computational tools, such as mobile devices, social media, programming, and robotic tangibles, provide various advantages for formal and informal learning (e.g., Dabbagh and Kitsantas 2012; Pereira, Baranauskas, and da Silva 2013; Khaddage, Müller, and Flintoff 2016; Burleson et al. 2018; Kjällander et al. 2018). It alters the way people learn as it allows novel and more instant methods in retrieving and generating knowledge through social interaction, new approaches of representation, and improved capability to cross space and time (Erstad and Sefton-Green 2013).

Guzdial (2015) had forecasted that a majority of students are likely to accomplish universal computing literacy through formal computing education pathways, namely elementary schools, secondary schools, and universities. Informal computing education (e.g., summer camps, online programs, MOOCs, museums, coding boot camps, and after school programs) is not appropriate for every student. Earlier studies from Fields, Giang, and Kafai (2014) and Ho et al. (2015) concluded that informal computing learning was biased toward more wealthy and male students compared to formal learning. Nevertheless, informal computing education still plays a crucial role in affecting identity and computing perception among the students. In another research study by Guzdial et al. (2014), it was found that informal learning in the summer camps has a significant effect on the pipeline of formal education. Studies by Ehsan et al. (2018) and Hynes et al. (2019) have proposed that students' CT can be assessed through formal and informal learning settings.

LEARNING PATHWAYS OF COMPUTING EDUCATION

Differentiated instruction is described as "an organization of pedagogical practices that promotes reporting learners' needs from a learning situation" (Tahiri, Bennani, and Khalidi Idrissi 2017, 200). It presents several advantages to the learners, such as giving learning content modified to the preferences of learners, including all the capabilities desired, showing the learner the most suitable learning path, and making learning activities and situations in a meaningful way. In other words, this instruction is applied based on examining the learners' diversity, considering them, and augmenting the learning conditions. Differentiated instruction is required in the classroom as the features of learners are heterogeneous and diverse in terms of education, skills, learning styles, needs, sociocultural backgrounds, learning profiles, level of engagement, and so forth. During instruction, multiple pedagogical methods, tools, and activities are employed to meet the learners' needs (Bermel 2016). This enables learners to become more motivated and productive (Tahiri, Bennani, and Khalidi Idrissi 2017).

In computing education, students possess varying levels of access to technology, and some of them may have disabilities in fully participating in computing activities. For example, students may have difficulty using the mouse such as not knowing how to left-click, double click, or drag an object. They may also face problems when using the keyboard functions such as "Control-Alt-Delete." Not only that, but they may also fail to read or interpret commands of Scratch and Etoys. Therefore, the

instructors have to utilize differentiated instruction including peer support, modeling, and scaffolding (Israel et al. 2015) to accommodate these
struggling students (Hansen et al. 2016). Through differentiated instruction, students can enter at their level and work to their potential in the
computing environment (Gadanidis and Caswell 2018).

According to Rajala et al. (2016), formal learning is regarded as planned
through academic institutions and leading to standard qualifications.
Voogt et al. (2015) claim that CT was positioned in the formal curriculum
in two major ways, namely computing as a separate subject and CT in
cross-curricular practices. It was argued by The Royal Society (2012) that
"every child should have the opportunity to learn concepts and principles from Computing (including Computer Science and Information
Technology) from the beginning of primary education onwards, and by
age 14 should be able to choose to study toward a recognized qualification in these areas" (44).

Informal learning is viewed as "what happens outside the structures
and boundaries of formal education, the topic or focus of which is determined by the person doing the learning, on their own or with others"
(Davies and Eynon 2013, 330). Wing (2008) contended that CT should
not just involve formal learning but also informal learning as "learning
takes place in many ways and outside the classroom: children teach each
other; learn from parents and family; learn at home, in museums and
in libraries; and learn through hobbies, surfing the Web and life experiences" (3721). Unlike formal learning, informal is tangible, open-ended,
interest- and practice-driven, and highly contextualized (Arnesen et al.
2016). Informal learning is not restricted by school culture (Wong, Jan,
and Liang 2014) as it occurs outside of prearranged lessons (McCartney
et al. 2010). It is conceived as learning that has less structure, intrinsic
motivation, and fewer formal assessments, and learning outcomes are not
evaluated explicitly. It is performed spontaneously and not always prearranged (Eshach 2006). Informal learning is voluntary, usually learner-led,
and nonsequential as compared to the rather compulsory, teacher-led, and
sequential formal learning (Eshach 2006). In both formal and informal CT
settings, the nature of the CT tasks and the supports and scaffolding given
by the adults will influence the engagement of the learners in different
levels of CT skills (Ehsan et al. 2018).

Leveraging computational tools has generated new potentials for linking learning taking place in different venues, linking people with mutual expertise and interests, and for incorporating informal learning within formal learning (Laru and Jarvela 2015). Digital practices in-school and out-of-school learning among the students is not separate, but it provides a supportive ecosystem for a varied range of students by making learning more engaging (Kumpulainen and Mikkola 2016).

The informal learning approaches introduced through inquiry-based, project-based, and problem-based learning techniques using technology would shift control from teachers to students (Schuck et al. 2017). Furthermore, informal practices and skills could be enhanced through digital making, mobile learning, social media, gaming, and getting involved in online communities. This would support the schoolwork in the formal classroom (Erstad, Gilje, and Arnseth 2013). In other words, daily digital practices can complement formal learning among young people.

Despite the disadvantages of informal education, it is believed that informal education can assist in reshaping formal education (Lewin and Charania 2018) by shifting the formal learning practices from transmissive techniques to student-centered, self-directed, and collaborative methods to deepen the knowledge of the students (Khaddage et al. 2016). A study was done by Boyle and McDougall (2003) to explore the formal and informal learning environments for the teaching and learning of computer programming. They found that although most of the learners are classified as formal learners with more external sources of control, informal learners show deep technical proficiency and various finished products, which shows that it may be best to leave some learners to obtain their own programming skills.

Boustedt et al. (2011) discovered many advantages of informal learning of computing topics from the computer science students including (a) the learner can select the level of study, either specific or conceptual; (b) a sense of achievement in informal learning, such as self-confidence and satisfaction; (c) things learned informally can usually be better engaged; (d) informal learning can be more flexible according to time and pace that suit the learner; (e) informal learning is motivating and can make topics interesting; (f) informal learning can be employed to increase the breadth or depth of learning in formal courses; and (g) informal learning can promote formal learning and vice versa. However, there are difficulties with informal learning, including (a) students may miss crucial features of the topic in informal learning; (b) informal learning is usually ad hoc and stops when the task can be solved using the knowledge gained so far; (c) some students need the structure and deadline provided by the teacher; (d) for learners, informal learning may be difficult to evaluate how learners know when they have learned enough knowledge; (e) without specific deadlines and rewards, less time may be spent on informal learning; (f) informal learners may miss chances for classroom discussions and other social interactions that are crucial for learning.

COMPUTATIONAL THINKING IN THE FORMAL CURRICULUM

There are two major trends seen in the manner in which various countries incorporate CT in the formal curriculum. The first trend is enhancing CT skills in children and young people to allow them to think differently, solve real-world problems, express themselves via various media, and investigate daily matters from a different viewpoint. The second trend is facilitating CT to augment the economic evolution, fill job openings in information and communication technologies, and prepare for a future career (Bocconi et al. 2016). The countries that have included CT into the formal curriculum can be classified into three clusters:

- cluster I: the ongoing curriculum reform process;
- cluster II: planning to introduce CT; and
- cluster III: building on a long-standing computer science education tradition.

Some countries have developed CT curriculum or policy initiatives at the regional level as shown in figure 7.1 (Bocconi et al. 2016). CT is positioned in the curriculum in terms of two criteria: (1) education level and (2) subject level.

Wing (2008) claimed that CT should not only be taught in the university but also incorporated into levels spanning primary to secondary. At the education level, countries are mostly seen to immerse CT at the secondary school level, but recently an emerging trend toward incorporation of CT at the primary school level has been seen as well. In some countries,

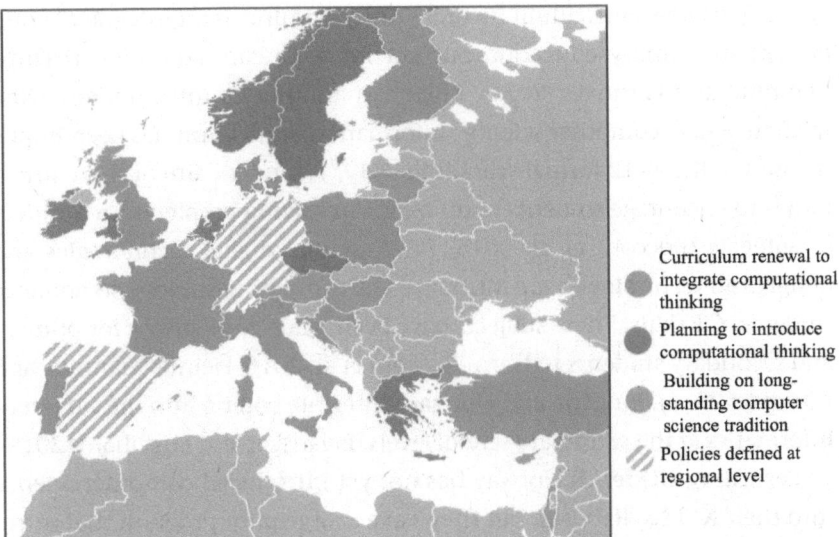

7.1 Integrating CT in the formal curriculum for European countries (Bocconi et al. 2016).

CT can be taught as a separate subject, and in others, it is embedded into subjects in the school syllabus (Voogt et al. 2015).

Among cluster I countries, Estonia introduced CT into the formal curriculum by launching the ProgeTiger program in 2012. This program intends to integrate robotics and programming into primary and secondary education. There are three thematic areas that are combined under this program, namely design and technology, ICT, and engineering sciences (Heintz, Mannila, and Färnqvist 2016). In Slovakia, coding is integrated as a compulsory component at all levels of school education. Therefore, all the students will learn coding throughout school (Balanskat and Engelhardt 2015).

England is one of the European countries that has introduced the computing subject in primary and secondary schools to replace the existing ICT curricula since 2014 (Bocconi et al. 2016). There are four key stages of K–12 education in England where students are anticipated to foster CT skills gradually, including key stage 1 (ages five to seven), key stage 2 (ages seven to eleven), key stage 3 (ages eleven to fourteen), and key stage 4 (ages fourteen to sixteen) (Department for Education 2019).

Another cluster I country is Finland. In 2016, Finland set up a new national curriculum for primary and secondary education (Mannila et al.

2014). This new curriculum includes programming as an integrated component in primary education but not for secondary education (Heintz, Mannila, and Färnqvist 2016). Poland is another country under cluster I, where a new computer science (informatics) curriculum has been implemented in the K–12 formal schools in 2017. The major aim of this curriculum is to encourage students to utilize CT in solving problems for a variety of subjects (Bocconi et al. 2016). In Australia, digital technologies and design and technologies are introduced as computing subjects to enhance students' CT skills. These subjects have been made compulsory for primary and secondary students to learn (Bocconi et al. 2016; Heintz, Mannila, and Färnqvist 2016). Bulgaria and Hungary integrate coding into the subject of informatics at the secondary school levels (Balanskat and Engelhardt 2015).

Regarding cluster II, Norway has not yet introduced computer science into their K–12 syllabuses, but they have a large pilot program that introduces programming as an elective subject at the secondary levels in 2016 (Heintz, Mannila, and Färnqvist 2016). In 2015, Sweden established a new curriculum for primary levels and also reorganized their secondary curriculum to reinforce the programming and digital competence. IT-related subjects are compulsory for primary students, but they are elective subjects for secondary students (Mannila et al. 2014).

For South Korea, a compulsory subject called "informatics" is introduced to the primary schools, and it is an elective subject in secondary schools (Heintz, Mannila, and Färnqvist 2016). Netherlands secondary schools have a computer science subject, but it is not a compulsory subject. The schools can decide whether to teach this subject to the students, and the students have the right to select this subject. In Ireland, an optional short course of coding has been introduced for the junior cycle program at the secondary school level, but there is no national program at the primary school level (Balanskat and Engelhardt 2015).

In cluster III, there are some countries that incorporated CT into formal education by building on a long-standing computer science education tradition. Israel is one of the countries that have had computer science education for a very long time at the secondary school level. Their computer science course aims to enhance students' algorithmic and logical thinking (Bocconi et al. 2016). Lithuania has a compulsory subject called "informatics" that has been revised to become "information technology" in secondary schools (Mannila et al. 2014). The five areas covered in information technology are digital technologies, virtual communication, information, programming, and algorithms in addition to ethics, security, and legal principles (Bocconi et al. 2016).

Several countries have developed CT curriculum or policy initiatives at the regional level. For instance, the state of Bavaria, Germany has had its own compulsory subject of computer science at the secondary levels since 2014. Digital competence, a programming-related subject, is mandatory for primary schools in one of the regions of Spain, Catalonia (Bocconi et al. 2016). Table 7.1 reveals the overview of computing education in different countries (adapted from Heintz, Mannila, and Färnqvist 2016).

COMPUTATIONAL THINKING IN THE INFORMAL CURRICULUM

According to Wing (2008), informal learning of CT ought to be explored as it happens anytime and everywhere. The students teach each other, learn from their family members, and learn in the libraries, in museums, and at home as well as through surfing websites and hobbies. Meetups like Codo Dojos, where students can learn programming from more experienced coders, and Code Clubs in after-school programs provide opportunities for learning in an informal environment. Fishman and Dede (2016) stated that the students can learn CT informally by engaging themselves as creators and makers, such as Do It Yourself digital textiles, Scratch programming, and robotics competitions. A study was done by McCartney et al. (2010) to determine how computing students learn computing topics informally. The findings revealed that the topics selected by the students were programming languages (e.g., C++ and Visual Basic), hardware (e.g., wiring two circuit boards), and software (e.g., Netbeans). These students chose to learn these topics because they desired to complete some projects. The resources utilized by the students were the Internet and thesauruses or books. Meanwhile, the learning approaches were searching for good examples, working with more knowledgeable mentors and friends, trial and error, talking to and observing others, and so on. The success of the projects was exploited to evaluate their learning. Overall, the informal learning setting brought a positive impact on the students' learning among computing students.

Table 7.1 Overview of computing education in different countries

Cluster	Country	Subject	Method	Primary	Secondary
I	Estonia	Programming (technology and innovation)	Integrated	Compulsory	Compulsory
I	Slovakia	Coding	Integrated	Compulsory	Compulsory
I	England	Computing	Replace existing subject	Compulsory	–
I	Finland	Programming (digital competence)	Integrated	Compulsory	–
I	Poland	Computer science (informatics)	Own subject	Compulsory	Compulsory
I	Australia	Digital technologies and design and technologies	Own subject and integrated	Compulsory	Compulsory
I	Bulgaria	Informatics	Integrated	–	Compulsory
I	Hungary	Informatics	Integrated	–	Compulsory
II	Norway	Programming	Own subject	–	Elective
II	Sweden	Programming and digital competence	Integrated	Compulsory	Elective
II	South Korea	Informatics	Own subject	Compulsory	Elective
II	Netherlands	Computer science	Own subject	–	Elective
II	Ireland	Coding (short course)	Own subject	–	Elective
III	Israel	Computer science	Own subject	–	Elective
III	Lithuania	Information technology	Revise existing subject	–	Compulsory

Lee et al. (2011) performed a study to enhance students' CT skills by conducting out-of-school programs through modeling and simulation, robotics, and game design and development. They engaged the youth in the rich CT environments using a three-stage progression, which was a use-modify-create framework. In the use stage, the students were required to use someone's creation, for instance, play a readymade computer game and run a program to control a robot. After that, the students would start to modify the game and program it to make it more sophisticated in the modify stage. Eventually, the students would create original products by employing their rising CT skills: automation, analysis, and abstraction.

Other work conducted by Ahmadi, Jazayeri, and Kandoni (2012) studied novice programmers in solving CT problems during the game design procedures in an informal learning context. AgentWeb, a cloud-based, online game design environment was utilized to allow the learners to construct their games beyond the formal classrooms. The learners' CT skills could be reinforced while identifying the game object, examining the behavior of the game object, and programming the behavior. Furthermore, Boulton et al. (2016) implemented a study to explore the role of game jams in stimulating CT among the learners aged eleven to eighteen in the informal learning setting. Game jams is an interesting approach that encourages a small game creation in a short time, with an aim to engage learners in learning computing and promoting their CT skills. The findings exhibited that the application of the mobile app, Pocket Code, in the game jams motivated the learners in the learning process.

Mesiti et al. (2019) researched how to develop CT capacity among youth through an informal learning setting in the Museum of Science, Boston. Three specific designs to convey CT content in the Pixar exhibition were utilized. Three dimensions of CT were focused on, namely concepts (i.e., patterns, decomposition, abstraction, and algorithms), practices (i.e., collaborating, creating, and debugging), and perspectives (i.e., empowered to ask questions about technology, sense of identity, and relationship to the technical world). The project research was accomplished in two phases. In phase 1, the focus was on how the learners could be reinforced to interact with exhibits and comprehend problem-solving approaches that address creative and complex challenges in computer programming. Phase 2 focused on exploring the affordances of these exhibits to develop CT capacity, interest, and feelings of efficiency in problem decomposition. The results displayed that the interest of youth in learning programming was augmented and their problem decomposition perception, creativity perception, and self-efficacy in computer programming became enhanced after the usage of exhibits. Not only that; they also understood the value and significance of mathematics and programming.

COMPUTING EDUCATION IN SINGAPORE

In 2014, Singapore launched the Smart Nation program, a nationwide effort to harness technology in sectors of business, government, and home to improve urban living, build stronger communities, grow the economy, and create opportunities for all residents to address the ever-changing global challenges (Smart Nation 2017). To support the Smart Nation initiative, one of the key enablers is to develop computational capabilities. Programs were implemented to introduce and develop CT skills and coding capabilities from preschool children to adults. We will next describe the landscape of K–12 CT- and coding-related programs in Singapore that are implemented by the schools and various government organizations.

Unlike countries like Finland, England, and South Korea, Singapore has not included computing or CT as compulsory education. However, learning coding has made inroads into schools initially through voluntary participation. Singapore's approach has been to provide opportunities for students to develop their interests in coding and computing skills through various touchpoints at various ages as shown in figure 7.2. Computing and CT skills are introduced to the children that are age-appropriate and intended to engage and stimulate their interests. For preschool and

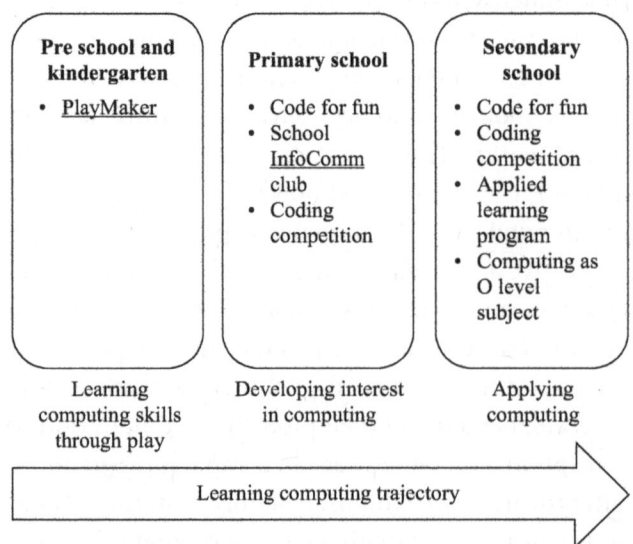

7.2 Learning pathways for computing education.

kindergarten children, the Playmaker program was introduced by the InfoComm and Digital Media Authority (IMDA), in collaboration with Temasek Polytechnic's Early Childhood Centre, to Singapore preschools (IMDA 2017). Teachers can use selected e-toys such as the BeeBot and Kibo to develop CT skills such as problem-solving and algorithmic thinking through play.

Many primary schools in Singapore opt-in to offer students the ten-hour or twenty-hour Code for Fun program conducted via the Ministry of Education (MOE) or by selected IMDA technology training partners where they are introduced to coding in Scratch and programming hardware such as Microbit or robotics kits. Lately in 2019, Code for Fun has been made compulsory for ten instructional hours during the upper primary years, that is, for grades 5 and 6 (Straits Times 2019). Secondary schools offer students opportunities to learn coding and develop CT skills through the Code for Fun, Digital Maker, and Applied Learning Programmes (ALP). As informal learning programs, co-curricular activities (CCAs) such as info-communications clubs in primary and secondary schools offer students opportunities to learn and apply computational skills. In the formal school curricula, computing is offered as a subject for grades 9 and 10 in the "O" levels, and computer science for grades 11 and 12 in the "A" levels. Thus, for students who take up these subjects, it becomes formal learning for them.

By incorporating both informal learning pursuits as well as formal learning courses, the programs provide varied experiences for our students to develop their own interests and apply computing to solving problems. The programs are facilitated, developed, and implemented by stakeholders in the ecosystem shown in figure 7.3. The agents in the ecosystem directly involved with students include teachers, school leaders, and parents. External agents include the government agencies such as the Infocomm Media Development Authority, Ministry of Education, Science Centre, and other computing educators. The range of programs for computing includes formal computing education in the O and A level of computing; informal programs in schools such as Code for Fun, Play/Digital Maker, computing-based CCAs, and the Applied Learning program; and the out-of-school enrichment programs such as the ones offered by computing educational centers or the Science Centre.

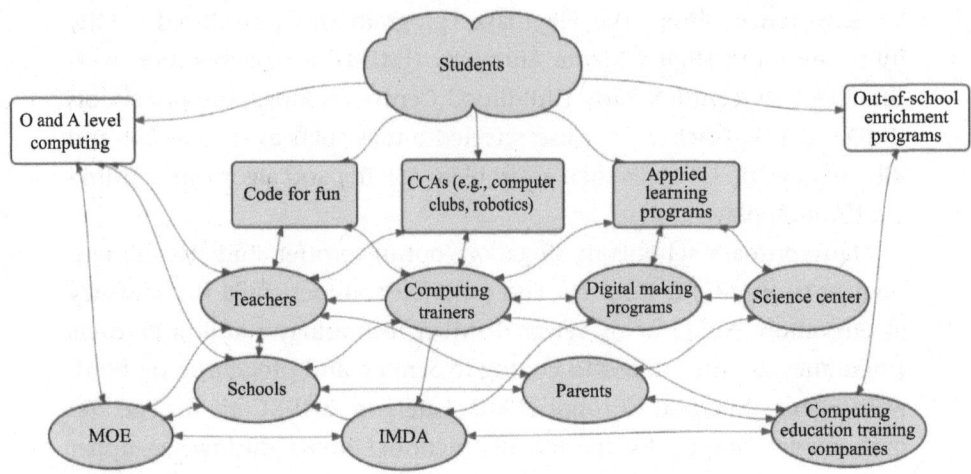

7.3   The computing education ecosystem in Singapore.

DISCUSSION AND CONCLUSION

In a nutshell, formal and informal systems play an important role as differentiated learning pathways in computing education. Both systems can be utilized to accommodate students of different levels and develop their CT skills in K–12 schools. The potential of informal learning to complement the formal learning about computing has been evidenced in earlier studies, such as Erstad, Gilje, and Arnseth (2013); Peeters et al. (2014); Guzdial et al. (2014); and Lewin and Charania (2018). However, the challenging part for informal learning of computing education is that it is not recognized as legitimate learning because it is implemented outside the school (Peeters et al. 2014). Also, immersing informal learning practices in the formal classroom remains problematic as a result of discrepancies with assessment practices, lack of technology infrastructure, and timetable limitations (Lewin and Charania 2018). Inadequate funds can also restrict out-of-school resources such as Internet and technology access, thus leading to inconsistencies in informal learning practices (Lopez and Caspe 2014).

A key aspect of addressing these challenges is professional development for teachers to implement CT in the classrooms. The computing or computer science teachers have to connect the core concepts of CT to other domains. Meanwhile, teachers from other domains should be familiar with the core concepts of CT (Voogt et al. 2015). The teachers need to be trained through professional development programs to fully harness informal learning to support the attainment of formal learning in computing education. They should provide more guidance and support to the students in connecting in-school and out-of-school learning for knowledge construction (Lewin and Charania 2018).

Hence, the suggestion is to infuse CT into the formal curriculum to make sure all children have equal chances to be equipped with the CT skills they require in a digital world. Not only that, but the teaching of CT should begin from early childhood as this can be a rewarding and engaging experience for young learners (Bers 2008). In earlier studies (e.g., Wyeth 2008; Kazakoff, Sullivan, and Bers 2012), children as young as four to six years old were able to construct and program simple robotics projects as well as build their CT skills and learn powerful ideas from computer programming, technology, and engineering (Bers 2008). Therefore, the government and policymakers ought to establish their goal and vision as well as judiciously describe, design, and observe their concrete execution actions. Setting particular targets is essential not only to notifying concrete execution options but also to getting related stakeholders on board (Bocconi et al. 2016). Thus, Bocconi et al. (2016) propose the policy and practice implications for introducing CT in formal education as shown in figure 7.4.

In figure 7.4, the policy and practice implications consist of various phrases (i.e., from preparation to execution), and they are all interconnected robustly. Four main parts should be the focus of policymakers and stakeholders: (1) consolidated CT understanding; (2) comprehensive integration; (3) systematic rollout; and (4) support policy. The first part, consolidating CT understanding, aims to promote a shared comprehension of CT and its relationship with twenty-first-century skills. It includes launching a shared comprehension on what CT is and how it is contextualized, elucidating differences and overlaps between CT and digital competence, and boosting grassroots initiatives to get involved in the policy discussion on CT.

The second part on comprehensive integration has three phases to infuse CT across all education levels: that is, enunciating a vision for

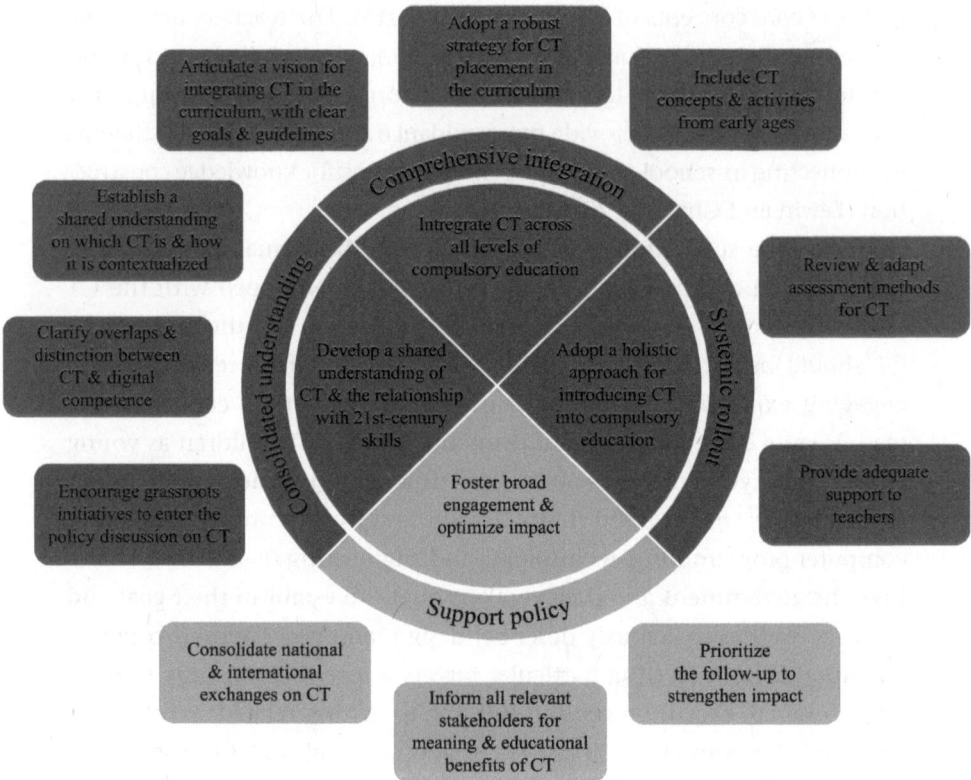

7.4 Policy and practice implications for introducing CT in formal education (Bocconi et al. 2016).

incorporating CT in the curriculum, utilizing a strong approach for CT positioning in the curriculum, and involving CT concepts and activities from early ages. Furthermore, there are two phases for systemic rollout to implement a holistic tactic for introducing CT, which are evaluating and modifying assessment techniques for CT, and giving sufficient support to teachers. Support policy, the fourth part of this framework, is intended to encourage extensive engagement and strengthen the effect of infusing CT across all education levels by combining national and international exchanges on CT, notifying all related stakeholders of the educational advantages and the meaning of CT, and prioritizing the follow-up to reinforce the impact (Bocconi et al. 2016).

ACKNOWLEDGMENTS

The work reported in this book chapter is supported by a grant from the Ministry of Education's Educational Research Funding Programme (OER 10/18 LCK).

REFERENCES

Ahmadi, Navid, Mehdi Jazayeri, and Monica Landoni. 2012. "Helping Novice Programmers to Bootstrap in the Cloud: Incorporating Support for Computational Thinking into the Game Design Process." In *Proceedings of the 2012 IEEE 12th International Conference on Advanced Learning Technologies, ICALT '12*, 349–353. Washington, DC: IEEE Computer Society.

Alves, Nathalia Da Cruz, Von Wangenheim, Christiane Gresse, and Jean C. R. Hauck. 2019. "Approaches to Assess Computational Thinking Competences Based on Code Analysis in K-12 Education: A Systematic Mapping Study." *Informatics in Education* 18, no. 1: 17–39.

Arnesen, Thomas, Eyvind Elstad, Gavriel Salomon, and Lars Vavik. 2016. "Educational Technology and Polycontextual Bridging: An Introduction." In *Educational Technology and Polycontextual Bridging*, edited by Eyvind Elstad, 3–14. Rotterdam/Boston/Taipei: Sense Publishers.

Balanskat, Anja, and Katja Engelhardt. 2015. *Computing Our Future: Computer Programming and Coding—Priorities, School Curricula and Initiatives across Europe*. Belgium: European Schoolnet.

Bermel, Jodi A. 2016. *Using One-to-one Computing for Differentiated Instruction in Iowa: An Investigation of the Impact of Teachers' Perceptions of Teaching and Learning*. Electronic Theses and Dissertations. 342.

Bers, Marina Umaschi. 2008. *Blocks, Robots and Computers: Learning about Technology in Early Childhood*. New York: Teacher's College Press.

Bocconi, Stefania, Augusto Chioccariello, Giuliana Dettori, Anusca Ferrari, and Katja Engelhardt. 2016. *Developing Computational Thinking in Compulsory Education*. Luxembourg: Publications Office of the European Union.

Boulton, Helen, Bernadette Spieler, Anja Petri, Christian Schindler, Wolfgang Slany, and Maria Beltran. 2016. "The Role of Game Jams in Developing Informal Learning of Computational Thinking: A Cross-European Case Study." In *Proceedings of EDULEARN16*, 4–6 July 2016, Barcelona, Spain, 7034–7044.

Boustedt, Jones, Anna Eckerdal, Robert McCartney, Kate Sanders, Lynda Thomas, and Carol Zander. 2011. "Students' Perceptions of the Differences between Formal and Informal Learning." In *Proceedings of the Seventh International Workshop on Computing Education Research*, 61–68. New York: ACM.

Boyle, Martin, and Anne McDougall. 2003. *Formal and Informal Environments for the Learning and Teaching of Computer Programming*. Paper presented at the IFIP Working

Groups 3.1 and 3.3 Working Conference: ICT and the Teacher of the Future, St. Hilda's College, The University of Melbourne, Australia 27–31 January 2003.

Burleson, Winslow S., Danielle B. Harlow, Katherine. J. Nilsen, Ken Perlin, Natalie Freed, Camilla Norgaard Jensen, Byron Lahey, Patrick Lu, and Kasia Muldner. 2018. "Active Learning Environments with Robotic Tangibles: Children's Physical and Virtual Spatial Programming Experiences." *IEEE Transactions on Learning Technologies* 11, no. 1: 96–106.

Computer Science Teachers Association (CSTA). 2016. "K–12 Computer Science Framework." Accessed October 2, 2019. http://www.k12cs.org.

Dabbagh, Nada, and Anastasia Kitsantas. 2012. "Personal Learning Environments, Social Media, and Self-Regulated Learning: A Natural Formula for Connecting Formal and Informal Learning." *Internet and Higher Education* 15, no. 1: 3–8.

Davies, Chris, and Rebecca Eynon. 2013. "Studies of the Internet in Learning and Education: Broadening the Disciplinary Landscape of Research." In *The Oxford Handbook of Internet Studies*, edited by William H. Dutton, 328–349. Oxford: Oxford University Press.

Department for Education. 2019. "National Curriculum in England: Computing Programmes of Study." Accessed October 3, 2019. https://www.gov.uk/government/publications/national-curriculum-in-englandcomputing-programmes-of-study/national-curriculum-in-england-computing-programmes-ofstudy.

Ehsan, Hoda, Tikyna Dandridge, Ibrahim H. Yeter, and Monica E. Cardella. 2018. "K-2 Students' Computational Thinking Engagement in Formal and Informal Learning Settings: A Case Study." In *Proceedings of the American Society for Engineering Education (ASEE) Conference & Exposition*. Salt Lake City, UT.

Erstad, Ola, and Julian Sefton-Green. 2013. "Digital Disconnect? The 'Digital Learner' and the School." In *Identity, Community, and Learning Lives in the Digital Age*, edited by Ola Erstad and Julian Sefton-Green, 87–104. New York: Cambridge University Press.

Erstad, Ola, Øystein Gilje, and Hans Christian Arnseth. 2013. "Learning Lives Connected: Digital Youth across School and Community Spaces." *Comunicar* 40: 89–98.

Eshach, Haim. 2006. "Bridging In-School and Out-Of-School Learning: Formal, Non-Formal, and Informal." In *Science Literacy in Primary Schools and Pre-Schools. Classics in Science Education* (vol. 1), edited by Haim Eshach, 115–141. Springer, Dordrecht. https://doi.org/10.1007/1-4020-4674-X_5.

Fields, Deborah. A., Michael Giang, and Yasmin Kafai. 2014. "Programming in the Wild: Trends in Youth Computational Participation in the Online Scratch Community." In *Proceedings of the 9th Workshop in Primary and Secondary Computing Education, WiPSCE '14*, 2–11, New York.

Fishman, Barry, and Chris Dede. 2016. "Teaching and Technology: New Tools for New Times." In *Handbook of Research on Teaching, 5th Edition (American Educational Research Association)*, edited by Drew H. Gitomer and Courtney A. Bell. New York: Springer.

Gadanidis, George, and Beverly Caswell. 2018. *Computational Modelling in Elementary Mathematics Education: Making Sense of Coding in Elementary Classrooms.* KNAER Mathematics Knowledge Network White Paper.

Guzdial, Mark. 2015. *Learner-Centered Design of Computing Education: Research on Computing for Everyone.* Morgan-Claypool.

Guzdial, Mark, Barbara Ericson, Tom Mcklin, and Shelly Engelman. 2014. "Georgia Computes! An Intervention in a US State, with Formal and Informal Education in a Policy Context." *ACM Transactions on Computing Education* 14, 13:1–13:29.

Hansen, Alexandria. K., Eric R. Hansen, Hilary A. Dwyer, Danielle B. Harlow, and Diana Franklin. 2016. "Differentiating for Diversity: Using Universal Design for Learning in Computer Science Education." In *Proceedings of the 47th ACM Technical Symposium on Computing Science Education—SIGCSE '16*, February 2016, 376–381.

Heintz, Fredrik, Linda Mannila, and Tommy Färnqvist. 2016. "A Review of Models for Introducing Computational Thinking, Computer Science and Computing in K-12 Education." *IEEE Frontiers in Education Conference (FIE)*, 1–9.

Ho, Andrew, Isaac Chuang, Justin Reich, Cody Coleman, Jacob Whitehill, Curtis Northcutt, Joseph Williams, John Hansen, Gienn Lopez, and Rebecca Petersen. 2015. "HarvardX and MITx: Two Years of Open Online Courses Fall 2012–Summer 2014." Accessed October 2, 2019. https://papers.ssrn.com/sol3/papers.cfm?abstract_id =2586847.

Hynes, Morgan M., Monica E. Cardella, Tamara J. Moore, Sean P. Brophy, Senay Purzer, Kristina Maruyama Tank, Muhsin Menekse, Ibrahim H. Yeter, and Huda Ehsan. 2019. "Inspiring Young Children to Engage in Computational Thinking In and Out of School." In *Proceeding of American Society for Engineering Education (ASEE) Conference & Exposition.* Tampa, FL.

IMDA. 2017. "PlayMaker Changing the Game." Accessed October 15, 2019. https:// www.imda.gov.sg/infocomm-and-media-news/buzz-central/2015/10/playmaker -changing-the-game.

Israel, Maya, Jamie N. Pearson, Tanya Tapia, Quentin M. Wherfel, and George Reese. 2015. "Supporting All Learners in School-Wide Computational Thinking: A Cross-Case Qualitative Analysis." *Computers & Education* 82: 263–279.

Kazakoff, Elizabeth R., Amanda Sullivan, and Marina U. Bers. 2012. "The Effect of a Classroom-Based Intensive Robotics and Programming Workshop on Sequencing Ability in Early Childhood." *Early Childhood Education Journal* 41, no. 4: 245–255.

Khaddage, Ferial, Wolfgang Müller, and Kim Flintoff. 2016. "Advancing Mobile Learning in Formal and Informal Settings via Mobile App Technology: Where to from Here, and How?" *Educational Technology & Society* 19, 3: 16–26.

Kjällander, Susanne, Anna Åkerfeldt, Linda Mannila, and Peter Parnes. 2018. "Makerspaces Across Settings: Didactic Design for Programming in Formal and Informal Teacher Education in the Nordic Countries." *Journal of Digital Learning in Teacher Education* 34, no. 1: 18–30.

Kumpulainen, Kristiina, and Anna Mikkola. 2016. "Toward Hybrid Learning: Educational Engagement and Learning in the Digital Age." In *Educational Technology and Polycontextual Bridging*, edited by Eyvind Elstad, 15–38. Rotterdam/Boston/Taipei: Sense Publishers.

Laru, Jari, and Sanna Järvelä. 2015. "Integrated Use of Multiple Social Software Tools and Face-to-Face Activities to Support Self-regulated Learning: A Case Study in a Higher Education Context." In *Seamless Learning in the Age of Mobile Connectivity*, edited by Lung-Hsiang Wong, Milrad Marcelo, and Specht Marcus, 471–484. Singapore: Springer.

Lee, Irene, Fred Martin, Jill Denner, Bob Coulter, Walter Allan, Jeri Erickson, Joyce Malyn-Smith, and Linda Werner. 2011. "Computational Thinking for Youth in Practice." *ACM Inroads* 2, no. 1: 33–37.

Lopez, M. Elena, and Margaret Caspe. 2014. *Family Engagement in Anywhere, Anytime Learning*. Cambridge, MA: Harvard Family Research Project.

Lunenburg, Fred C. 2010. Extracurricular Activities. *Schooling* 1, no. 1: 1–4.

Manches, Andrew, and Lydia Plowman. 2017. "Computing Education in Children's Early Years: A Call for Debate." *British Journal of Educational Technology* 48, no. 1: 191–201.

Mannila, Linda, Valentina Dagiene, Barbara Demo, Natasa Grgurina, Claudio Mirolo, Lennart Rolandsson, and Amber Settle. 2014. "Computational Thinking in K-9 Education." In *Proceedings of the Working Group Reports of the 2014 on Innovation & Technology in Computer Science Education Conference, ITiCSE-WGR 2014*, 1–29. New York: ACM.

McCartney, Robert, Anna Eckerdal, Jam Erik Moström, Kate Sanders, Lynda Thomas, and Carol Zander. 2010. "Computing Students Learning Computing Informally." In *Proceedings of the 10th Koli Calling International Conference on Computing Education Research—Koli Calling '10*, 43–8.

Mesiti, Leigh Ann, Alana Parkes, Sunewan C. Paneto, and Clara Cahill. 2019. "Building Capacity for Computational Thinking in Youth through Informal Education." *Journal of Museum Education* 44, no. 1: 108–21.

Peeters, Jeltsen, Free De Backer, Tine Buffel, Ankelien Kindekens, Katrien Struyven, Chang Zhu, and Koen Lombaerts. 2014. "Adult Learners' Informal Learning Experiences in Formal Education Setting." *Journal of Adult Development* 21, no. 3: 181–192.

Pereira, Roberto, M. Cecilia C. Baranauskas, and Sergio Roberto P. da Silva. 2013. "Social Software and Educational Technology: Informal, Formal and Technical Values." *Educational Technology & Society* 16, 1: 4–14.

Rajala, Antti, Kristiina Kumpulainen, Jaakko Hilppö, Maiju Paananen, and Lasse Lipponen. 2016. "Connecting Learning across School and Out-of-School Contexts: A Review of Pedagogical Approaches." In *Learning across Contexts in the Knowledge Society*, edited by Ola Erstad, Kristiina Kumpulainen, Asa Makitalo, Kim Christian

Schroder, Pille Pruulmann-Vengerfeldt, and Thuridur Johannsdottir, 15–38. Rotterdam/Boston/Taipei: Sense Publishers.

Roman-Gonzalez, Marcos, Juan-Carlos C. Perez-Gonzalez, Jesus Moreno-Leon, and Gregorio Robles. 2018. "Can Computational Talent Be Detected? Predictive Validity of the Computational Thinking Test." *International Journal of Child-Computer Interaction* 18: 47–58.

Schuck, Sandy, Matthew Kearney, and Kevin Burden. 2017. "Exploring Mobile Learning in the Third Space." *Technology, Pedagogy and Education* 26, no. 2: 121–137.

Smart Nation. 2017. "Why Smart Nation." Accessed October 10, 2019. https://www.smartnation.sg/about-smart-nation.

Snodgrass, Melinda R., Maya Israel, and George C. Reese. 2016. "Instructional Supports for Students with Disabilities in K-5 Computing: Findings from a Cross-Case Analysis." *Computers & Education* 100: 1–17.

Strait Times (2019). "Coding Classes for All Upper Primary Pupils from 2020." Accessed July 12, 2019. https://www.straitstimes.com/tech/coding-classes-for-all-upper-primary-pupils-from-2020.

Tahiri, Jihane Sophia, Samir Bennani, and Mohamed Khalidi Idrissi. 2017. "diffMOOC: Differentiated Learning Paths Through the Use of Differentiated Instruction within MOOC." *International Journal of Emerging Technologies in Learning (iJET)* 12, no. 03: 197–218.

The Royal Society. 2012. "Shut Down or Restart? The Way Forward for Computing in UK Schools." London: The Royal Society. Accessed October 9, 2019. http://royalsociety.org/uploadedFiles/Royal_Society_Content/education/policy/computing-in-schools/2012-01-12-Computing-in-Schools.pdf.

Voogt, Joke, Petra Fisser, Jon Good, Punya Mishra, and Aman Yadav. 2015. "Computational Thinking in Compulsory Education: Towards an Agenda for Research and Practice." *Education and Information Technologies* 20, no. 4: 715–728.

Wing, Jeannette M. 2006. "Computational Thinking." *Communications of the ACM* 49, no. 3: 33–36.

Wing, Jeannette. 2008. "Computational Thinking and Thinking about Computing." *Philosophical Transactions of the Royal Society* 366, no. 1881: 3717–3725.

Wong, Lung Hsiang, Mingfong Jan, and Rose Liang. 2014. "Learning Sciences". In NIE Working Paper Series No. 1, edited by Wing On Lee and David Hung, 5–41. Office of Education Research, National Institute of Education, Nanyang Technological University.

Wyeth, Peta. 2008. "How Young Children Learn to Program with Sensor, Action, and Logic Blocks." *International Journal of the Learning Sciences* 17, no. 4: 517–550.

8

SOFTWARE EDUCATION IN SOUTH KOREA FOR CULTIVATING COMPUTATIONAL THINKING: OPPORTUNITIES AND CHALLENGES

Hyo-Jeong So, Dongsim Kim, and Dahyeon Ryoo

INTRODUCTION

In the intelligent information society brought by the fourth Industrial Revolution, software-based innovations have become the center of value creations, enabling the design of autonomous physical-digital spaces and intelligent systems that change the way we live, learn, and work. Accordingly, several countries have been emphasizing the importance of equipping the future workforce with necessary computational skills, creativity, and logical thinking skills (Hsu, Chang, and Hung 2018). As a strategic policy move under the title of "software education"[1] (SW education hereinafter), the South Korean government has mandated SW education in elementary and middle schools. SW education, which is defined as "the education of the ways of thinking that enable students to express creative ideas through software" (Ministry of Education in South Korea 2015), was introduced in 2014 as part of the policy strategy to support a software-centered society. The educational intention of this policy initiative is to foster the culture of learning that highlights problem-solving and computational thinking (CT) skills.

With this backdrop of the education reform, the purpose of this chapter is to discuss the current status, opportunities, and challenges of implementing software education in South Korea. From 2018, software education has been introduced and mandated in the school curricula to cultivate students' computational thinking and related competencies such as creativity and communication skills. In this chapter, we first present the current status of software education in K–12 education with a particular focus on the unique features of software education in South Korea, namely (a) the inclusion of software education in the formal curricula as a single subject, (b) the top-down policy approach, and (c) the integration with STEAM (science, technology, engineering, arts, and mathematics) education. We then discuss the difficulty of changing teachers' perceptions of viewing software education as computational thinking beyond coding skills. The chapter concludes with the discussion of both opportunities and challenges that K–12 schools in South Korea have been facing in terms of changing the culture of learning coding skills, the need for developing teacher competency in both technical and pedagogical aspects, and the emergence of the commercial market of coding education.

POLICY DIRECTIONS AND INITIATIVES

INTRODUCTION OF THE NEW CURRICULA

The national-level or school-level curriculum in SW education has been determined based on the Elementary and Secondary Education Act. As summarized in figure 8.1, elementary school students receive SW education in the practical study subject for seventeen hours per year in the fifth

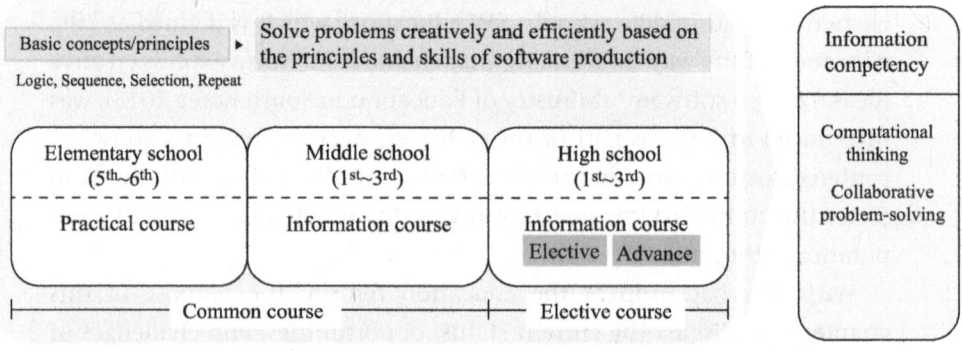

8.1   The SW education curriculum structure for each school level in South Korea.

and sixth grades. In the middle school curriculum, students receive SW education in the information subject for thirty-four hours per year.

While total curricula hours allocated for SW education are not high, the national policy on SW education has brought several changes in the existing curricula. First, the focus of SW education has changed from learning software applications to coding education. In elementary schools, while the previous curriculum centered on ICT literacy skills such as using and developing multimedia materials, the new curriculum called the 2015 National Curricula emphasizes diverse topics around computational thinking such as algorithms, information ethics, and programming. In middle schools, the previous curriculum in information science focused on teaching computer literacy skills such as creating documents and editing video clips, whereas the new curriculum includes programming skills and various topics related to computational thinking. Specifically, the components of the revised curriculum include information culture–information society–information ethics (18 percent); data and information—representation and analysis (18 percent); problem-solving and programming—abstraction, algorithms, and programming (50 percent); and the operating principles of computing systems and physical computing (14 percent). As seen in the distribution of topics, the new middle school curriculum puts a high emphasis on developing students' competency in computational thinking.

In high schools, while information science was offered as an advanced elective, most schools did not offer this subject due to the low relevance to the college entrance exam. Since the revised curriculum has changed the information subject from an advanced elective to a general elective under the science and technology subject, as of 2019, about 10 percent of high schools offer the information subject under the formal curriculum. The information subject has also been redesigned to cover various topics such as information ethics, information security, computer logic gates, the basics of C language, advanced concepts (pointers, structures, and so on), algorithms, data structures, and alignment algorithms. In particular, the new subject called advanced information science has been developed to allow students who are interested in the information science field to learn and develop advanced skills through the sustainable curriculum (see figure 8.2).

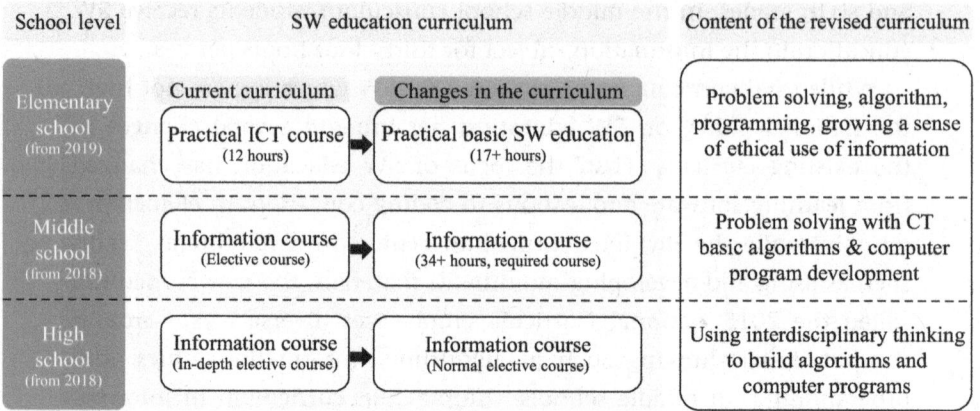

8.2   Changes in SW education in the 2015 national curricula.

Second, the introduction of SW education requires teachers to develop new competencies to provide SW education that is different from the past curriculum. While middle school teachers have advanced content knowledge in SW education, elementary school teachers are trained to teach various subjects and do not have the specific content knowledge and technical skills to teach computational thinking. Recognizing the lack of elementary school teachers' competency with SW education, the Ministry of Education established the strategic plan to offer professional development programs to 30 percent of elementary school teachers by 2019. Teacher professional development programs under the SW education initiative include software capacity building training through online learning (basic and advanced), face-to-face training, group training for lead teachers, and school-organized workshops. In addition, there are several teacher study groups formed in a voluntary way.

Third, there has been considerable investment to construct learning culture and environments conducive to SW education. Since technology is both an objective and a means in SW education, it is impossible to implement SW education in schools without creating a sufficient physical and technical environment (So, Kim, and Ryoo 2020). To change the school culture along with the support of teacher professional development, the number of lead schools in SW education has increased dramatically from 160 in 2015 to 1832 in 2019. SW education lead schools receive financial support from the Ministry of Education, which can be used for purchasing teaching and learning materials and organizing SW training camps, teacher workshops, and parent seminars. Further, wireless infrastructure in schools has been improved along with the supply of smart devices and computers.

INTEGRATION WITH STEAM EDUCATION

Through SW education offered in the information subject, students are exposed to a multidisciplinary education that emphasizes solving real-world problems. SW education aims to help students integrate various types of knowledge for problem-solving to go beyond learning separate subjects. For this purpose, SW education has been expanded with the integration of STEAM education and maker education. SW education can be used as part of STEAM education in which various subjects are integrated to solve challenging problems. In view of this, a new approach called computational thinking–based STEAM (CT-STEAM), which converges multiple subject knowledge to solve complex problems in the real world by utilizing various computing devices, has also been proposed (Ham, Kim, and Song 2014).

STEAM education also includes unplugged computing, which is an activity for learning about computer science concepts without the use of computers (Bell et al. 2009). Unplugged activities in the STEAM curriculum have been provided in the formal curriculum for young learners and primary school students to increase their interest in learning about computer science concepts. Unplugged computing has attracted much attention as a new method in the STEAM curriculum because the logical aspects of computer science can be taught without complex programming (Thies and Vahrenhold 2013). For upper-level elementary and middle school students who are interested in computer science, STEAM education focuses on helping them solve complex problems through programming robots and computing applications with the integration of engineering, coding, sciences, and mathematics subjects (Ngamkajornwiwat et al. 2017; Sullivan and Bers 2016). Students naturally learn about the functions and roles of robots by playing with them (Jeon et al. 2016). STEAM enrichment programs using robots can enhance students' creativity and complex problem-solving abilities (Ngamkajornwiwat et al. 2017). In South Korea, there are research studies on the STEAM enrichment programs using Littlebits in design thinking approaches (Tae 2016) and drones and hamster robots (Yoo et al. 2018).

Entry (see figure 8.3) is the most commonly used coding program in SW education in South Korea. Entry has been used in most elementary schools, whereas about 50 percent of middle and high schools use Entry. In 2013, Entry was developed by NAVER, which is the most popular web portal site in South Korea. and is a block-based coding program for users to easily learn coding skills. Entry is supported by educational materials developed with in-service teachers and experts and provides teachers with grade-level specific teaching and learning materials based on the national curriculum and lesson plans developed in collaboration with EBS (Educational Broadcasting System), which is a public broadcasting organization in South Korea. Entry also supports unplugged activities such as board games and physical computing activities (see figure 8.4).

THE EFFECTS OF SW EDUCATION

As SW education has been fully introduced in the formal curricula in schools, several research studies have reported the effects of SW education on various learning outcomes. In this section, we synthesize research

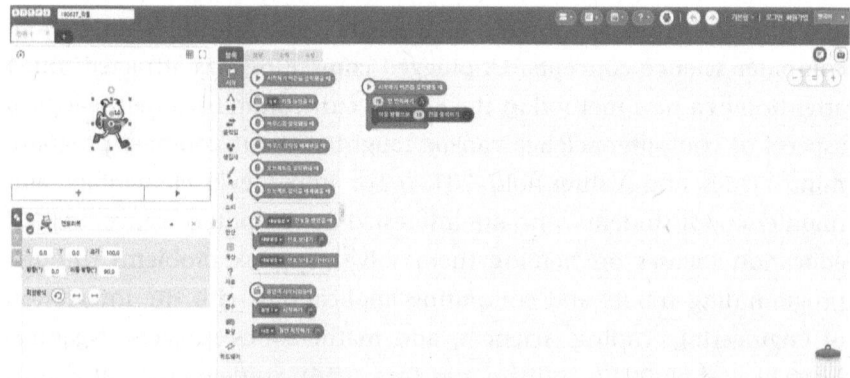

8.3 Entry interface. (image source: https://playentry.org) Copyright © Naver Connect Foundation. Some rights reserved.

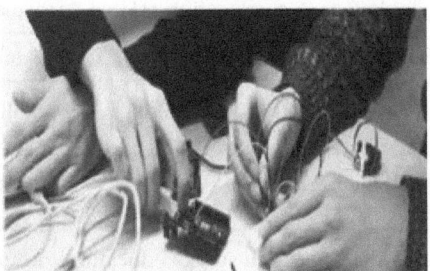

8.4 Unplugged activity (left) and physical computing activity (right) in Entry. (image source: https://playentry.org) Copyright © Naver Connect Foundation. Some rights reserved.

studies that examined the effect of SW education in South Korea, according to each school level from early childhood education to high schools. For the literature search, we used the RISS service, which is the most widely used academic search database in South Korea. As shown in table 8.1, the education system in South Korea is divided into elementary, middle, and high schools, and SW education has been offered differently at each education level. As mentioned earlier, SW education has been mandated in elementary, middle, and high school as part of the formal curricula. The Nuri curriculum is commonly applied to kindergartens and daycare centers across the country but does not include specific components for SW education. With increasing interest in coding education with the emphasis on playful activities, many kindergartens have been implementing coding education programs.

First, research on SW education in early childhood education has been conducted as part of the Nuri curriculum to examine children's attitudes and perceptions about software applications. Jung and Park (2018) examined the effect of using unplugged activities for children. The comparison between the control group ($N=18$) and the experimental group ($N=20$) showed that children in the experimental group that used unplugged activities showed higher levels of creativity and problem-solving than their counterparts in the control group. In general, children understand the process of building and operating robots through playful activities. With that, recent programs attempt to integrate playful activities in SW education. For instance, Kim, Hong, and Kim (2016) developed

Table 8.1 SW education-related curricula at each school level

School level	Educational target	Related curriculum
Early childhood education	Children before primary education (aged 0–7)	Nuri curriculum
Elementary school education (6 years)	Students aged 8–13	SW education, STEAM education
Middle and high school education (3 years+3 years)	Middle school: students aged 14–16; high school: students aged 17–19	SW education, STEAM education, free semester

instructional tools with QR codes to teach CT through playful activities. To do this, they categorized children's CT into ten categories such as transformational thinking, recursive thinking, and abstract thinking and developed relevant teaching materials in language, manipulation, and rhythm areas. Some gaps in the existing research with young children's CT were also found. While the existing literature examined the impact of SW education for young learners, little is known about how children form CT in the developmental process, implying the need for shifting the research focus from outcome-oriented to process-oriented approaches.

Second, several research studies conducted in the elementary education context examined the effects of the SW education activities such as block coding programs (e.g., Scratch, Entry, and Bitbrick) and hamster robots. Such learning activities helped students develop CT-based problem-solving skills and logical thinking. For instance, Lee, Park, and Choi (2018) conducted the SW education program with robots for eighty-eight students in the fifth grade for three months and found that students' CT, creativity, and academic interests improved significantly after participating in the program. Some studies examined the effectiveness of SW education integrated with the STEAM approach. Kim (2015) developed the 4C (creativity, communication, caring, and convergence)-STEAM education model with twenty-two curriculum activities under six themes in engineering and technology. The study reported that the group who participated in the STEAM curriculum was more satisfied with the learning experience and showed higher logical thinking ability than the control group.

In addition, some studies examined the different levels of CT according to learner characteristics. For example, Park et al. (2017) investigated 4,363 students in the fifth and sixth grades in forty-five SW leading schools to identify the main factors affecting students' CT and attitude toward SW education. Their study examined three levels of student characteristics factors: level 1—within-student factors (i.e., CT and attitude toward SW education); level 2—between-student factors (i.e., gender, computer owner-ship, Internet usage time, prior SW education experiences, participation in SW competitions, awareness of SW education, satisfaction with SW edu-cation, time on SW training, and participation in SW education through in-school and after-school programs); and level 3—school-level factors (i.e., city size and the number of students). The study found that the period of SW education and participation in the SW-related after-school programs at level 2 had the greatest influence on students' CT. Other level 2 variables such as the awareness of SW education and satisfaction with SW educa-tion showed significant influences on students' attitudes toward SW edu-cation. As another related study, Noh and Lee (2017) examined the SW education program with robotics for 155 students in the fifth and sixth grades and analyzed their level of CT and creativity according to prior knowledge and gender. They found that prior knowledge related to SW education played a significant role in CT and suggested the need to sup-port programs that CT could be improved through various experiences.

Third, research conducted in the middle and high school context is relatively scarce compared to the number of studies done in the elemen-tary school context, and most existing studies focus on identifying factors affecting students' CT. For instance, Lee and Ko (2018) examined the level of CT among eighty-three students in middle school who received SW edu-cation in the information subject for six months. They found that the sub-factors of CT (i.e., critical thinking, creativity, algorithmic thinking, and problem-solving ability) increased significantly, and the students acquired CT-related concepts and perspectives through the SW education activi-ties. Hwang, Mun, and Park (2016) examined the impact of unplugged activities in the SW education club for twenty-three high school students who used education programming language (EPL) software and physical computing tools. The study found that the students showed improved programming skills, CT, and confidence in problem-solving through computing after participating in the club activities. Additionally, some studies examined differences in student perceptions about CT and SW according to individual characteristics. Lee, Jo, and Kim (2019) examined 422 middle school students' awareness about SW (i.e., SW environment, SW education awareness, and SW career interest) according to their gender. They found that the male students were more interested in learning about SW than the female students. In addition, while 76 percent of the male students were exposed to SW education, only 54 percent of the female students had SW-related prior experiences. Overall, they suggested the need to increase female students' confidence and interest toward SW education.

In summary, our review reveals that most studies in the area of SW education in the South Korean context have been conducted with students in elementary schools and have attempted to identify factors related to the effects of SW education. Given the increase of research on young children, we also confirmed the expansion of SW education across various school levels. Kindergarten and elementary school students tended to use unplugging activities and block coding, whereas middle and high school students were participating in higher-level SW classes with Python and C programing language. For the sustainability of SW education, it needs to be integrated with other subject studies, as seen in the increased interest in integrating SW education with STEAM approaches. In addition, we also suggest the need for longitudinal studies of cohorts at each school level to examine the long-term outcomes of SW education.

BARRIERS TO SW EDUCATION

While teachers play an important role in the spread and adoption of SW education in South Korea, they also experience some barriers in implementing SW education because of the difficulty in applying new teaching methods and technologies. Teachers need to acquire certain content knowledge in computer science to competently teach SW education in the classroom (Saeli et al. 2011). Teachers also need to design challenging tasks and provide pedagogical support to help students solve computational problems that are connected to authentic contexts (Krauss and Prottsman 2017).

Ertmer (1999; 2005) suggests that teachers face both external (first-order) and internal (second-order) barriers when integrating new technologies into teaching and learning practices. First-order barriers refer to obstacles that are external to teachers and are often described as a lack of resources such as insufficient equipment, time, financial, and administrative support. These types of barriers can be overcome incrementally at an institutional level once adequate resources are provided. On the other hand, second-order barriers are intrinsic to teachers such as epistemological beliefs and attitudes. Ertmer (1999; 2005) argues that second-order barriers are more difficult to change or remove as they are deeply ingrained, intangible, and personal factors. Teachers are reluctant to use new technology when they perceive a high degree of uncertainty and risk that new technology brings along (Le Fevre 2014). To overcome second-order barriers, it is necessary to change teachers' belief systems by providing opportunities to experience the potential of new pedagogical methods (Blackwell et al. 2013; Donnelly, McGarr, and O'Reilly 2011). In this section, we discuss various barriers that South Korean teachers have experienced in the implementation of SW education in terms of first-order and second-order barriers.

Some research studies have reported environmental and structural factors as first-order barriers. Kim and Jun (2019) conducted a study with 181 elementary school teachers to identify their perceptions about the SW education policy. The IPA (Importance-Performance Analysis) revealed that teachers perceived policy support and skill enhancement through professional development as important factors. There was a great demand for policy support for improving school environments, such as securing digital devices, wireless networks, and budget support. Teachers also perceived high importance about learning instructional strategies that can be applied to SW lessons, such as analyzing learners, accommodating learners' diverse ability levels, and utilization ICT tools. The study by Ryoo, So, and Kim (2019) demonstrates how first-order barriers can affect second-order barriers such as teachers' beliefs. Their study classified four types of teachers, namely the innovative teacher, the serious teacher, the insecure teacher, and the confused teacher, based on knowledge/skill levels (surplus versus shortage) and pedagogical beliefs (positive versus negative). Teachers commonly pointed out the lack of resources in SW

classes as one of the serious factors affecting their adoption and implementation of SW education. In particular, the serious teacher (surplus of
skills, low pedagogical belief), despite having sufficient knowledge and
skills, experienced difficulty in operating SW education due to the insufficient financial support and lack of school infrastructure. On the other
hand, the insecure teacher (shortage of skills, high pedagogical belief) did
not have enough knowledge and skills to implement SW education and
expressed concerns about how to implement SW lessons in class. Despite
their concern about the lack of knowledge and skills, an encouraging finding was that the insecure teacher type tends to believe that SW education
can provide students with opportunities to develop necessary competence
demanded in the future society. The Ministry of Education in South Korea
has recognized such problems related to school environments that are
not conducive to SW education and has established a strategic plan to
provide wireless Internet infrastructure to elementary schools by 2020
and middle schools by 2021 and to replace outdated hardware applications and devices.

Another problem related to the first-order barriers is that teachers did
not receive enough training about SW education during their teacher education programs. Since the SW curriculum is mandatory and requires
curriculum hours, teachers need to develop teaching competence to
design, implement, and evaluate the mandatory curriculum implementation. To overcome this barrier related to teacher training, the Ministry of
Education and the Ministry of Science and Technology and Information
and Communication in South Korea have provided training programs to
6,800 teachers (30 percent of all elementary school teachers) and about
1,800 information subject teachers in middle schools. In particular, one
of the national tasks is that the core teacher training in SW education
will be conducted until 2021, with the aim to cultivate ten thousand
core teachers leading SW education (Ministry of Education in South Korea
2018). Teachers in elementary schools will receive training courses about
understanding CT, unplugged activities, acquisition of EPL, and SW
education with robots. Teachers in middle and high schools will receive
training courses on more advanced topics such as physical computing
and text-based programming (Korea Education & Research Information
Service 2017a, 2017b).

In addition, there are attempts to hire assistant teachers for in-service teachers who do not have enough knowledge and skills of programming and have great difficulty in the teaching process. The use of assistant teachers in South Korea has been already implemented in English education and inclusive education for children with disabilities. Noh (2018) examined the impact of having assistant teachers in SW classes. Both the experimental group and the control group had the same teacher and class contents, but the assistant teacher in the experimental group was responsible for observing students, answering questions, and providing necessary information in the debugging process. The study found that students in the experimental group performed significantly better than their counterparts in the control group. Besides human support, teachers need guidelines and educational materials for SW education. Hence, the Ministry of Education, the Ministry of Science and Technology, and various institutions such as KERIS and EBS have collaborated to develop teaching and learning materials for SW education.

Next, previous studies have identified second-order barriers such as teachers' perceptions and attitudes about SW education. Lee, Chung, and Ko (2018) identified factors that influenced elementary school teachers' intention to implement SW education based on the Technology Acceptance Model (TAM), which explains individuals' intention to use new technology through two factors, namely the perceived ease of use and the perceived usefulness (Davis 1989). Lee et al. (2018) found that teachers' attitudes about SW education completely mediate their perceived ease of use, perceived usefulness, personal innovation, and intention to implement SW education. In addition, the perceived ease of use had a direct effect on the perceived usefulness and an indirect effect on the intention to implement SW education.

Lee and Lee (2019) examined the variables that predict teachers' beliefs about SW education. Training experience and time related to professional development about SW education were predictive variables. In another study, they examined the impact of PD programs about backward design on teachers' beliefs. The training program provided the teachers with the opportunity for in-depth experiences designing lesson plans based on the backward design approach. It found that the process of professional development was improved by teachers' beliefs. In summary, a teacher's attitude about SW education was a major factor in predicting their intention to implement SW education, implying that teachers' positive attitudes influence not only the development of teacher expertise in SW education but also actual classroom implementations.

However, it is important to note that acquiring the necessary knowledge alone does not help teachers develop sufficient professional competency. From the perspective of technological pedagogical content knowledge (TPCK) proposed by Koehler and Mishra (2009), So and Kim (2009) argue that even when teachers have strong beliefs about technology integration (espoused TPCK), they tend to experience difficulty designing technology-integrated lessons in practices (in-use TPCK). This problem arises when teachers have not formed tight connections among pedagogy, content, and technology. One way to overcome this conflict between espoused TPCK and in-use TPCK is to help teachers acquire explicit knowledge and change beliefs at the same time (Kane, Sandretto, and Heath 2002). Belief is a source of change for teachers, and a belief system is formed over time through embodied experiences (Hansen and Rosenlund 2018). Hence, teachers need to develop relevant knowledge, skills, and beliefs to innovate teaching and learning practices with new technologies. For instance, Angeli et al. (2016) propose the concept of TPCK for computational thinking ($TPCK_{CT}$) that refers to teacher knowledge in terms of identifying a range of creative and authentic projects for CT, identifying a range of technologies to support CT in each project, and integrating content knowledge and pedagogical knowledge to make the overall CT experiences comprehensive to all learners. Angeli et al. (2016) also reported that the learning-by-design approach where teachers are engaged in designing models of different problem situations and constructing computer programs could improve $TPCK_{CT}$. A similar approach emphasizing teachers as designers can be adopted in the teacher preparation program in South Korea to equip future teachers to better deal with the complexity of designing and implementing the SW education curriculum and to help their students develop adequate CT.

DISCUSSION AND CONCLUSION

The increasing interest toward CT in K–12 education is a global phenomenon, and many countries have initiatives to promote CT in formal education. South Korea is not an exception in this global CT movement. The SW education initiative in K–12 schools in South Korea has been positioned as the major education reform movement toward cultivating students' computational skills and creative problem-solving skills demanded in a creative economy. In tracing the historical development and spread of CT, Tedre and Denning (2016) argue that translating ideas underlying CT into K–12 education is a major challenging as most teachers lack computer science knowledge. Further, they list several threats to CT initiatives, namely the lack of ambition, dogmatism, knowing versus doing, exaggerated claims, narrow views of computing, overemphasis on the formulation, and lost sight of computational models. This chapter also reveals that cultivating CT in South Korean K–12 schools under the overarching SW education policy has faced several challenges despite the macro-level support from the government. Based on the review of the current status and research on SW education, we have drawn the following four features about the early stage of SW education in South Korea.

First, SW education in South Korea has been introduced based on the national curriculum. Although there were some difficulties in the initial process of its introduction, the awareness of the necessity of introducing SW education has been increasing gradually. Teacher changes have been observed in teacher learning communities where they attempt to build necessary knowledge and skills through bottom-up initiatives. For example, the Association of Teachers for Computing, which is the largest teacher-initiated community for SW education in South Korea, has various channels such as social media and workshops to communicate with teachers and has made various teaching and learning materials available for teachers. When there is a mechanism to support teachers' voluntary and bottom-up movements such as this teacher community, SW education can be successfully promoted and implemented in schools.

Second, there have been continuous efforts for improving teachers' SW teaching competency through the provision of various professional development programs. There are high demands on offering differentiated training programs according to individual teacher's prior knowledge, gender, teaching experiences, and training experiences (Ryoo et al. 2019). As of the 2019 statistics, the average age of teachers in South Korea is about 40.4 years old for elementary schools, 43 years old in middle schools, and 43.3 years old in high schools (Korea Educational Development Institute

2019). Since teachers' age has been reported as one of the important variables in technology integration, the design of teacher training programs for SW education needs to consider teachers' life cycles (Kale and Goh 2014).

Third, regional differences in implementing SW education have been reported, indicating the educational disparity. In South Korea, as technology infrastructure has been built around urban areas, less manpower and equipment have been provided in rural areas. The proportion of information science teachers who could teach SW education was 49.9 percent in 2019. Itinerant teachers who travel to teach at various schools in the region are used for solving the problem of lack of qualified SW education teachers in rural areas. Further, standardized lesson plans and necessary materials have been developed to help teachers in rural areas who have fewer technical skills implement lessons in SW education.

Lastly, the emphasis on SW education in the formal curriculum has led to the expansion of the private education market for coding classes, which has been growing dramatically without relevant regulations. The private education market for coding education has appeared in various formats such as coding schools, coding cafés, and after-school programs. It has been reported that about 84.5 percent of students received coding education through private education institutions (Hur 2019). Such a large private education market is a serious problem as it may create another education gap between those who can afford the cost of private education and those who cannot afford it. Since private education institutions are centered in urban areas, this phenomenon may become another source of regional differences in the quality of education. Further, since private tutoring focuses on producing visible results quickly, it can obscure the essence of SW education, which is to enhance CT skills. However, considering that private tutoring has the flexibility to respond to changes in technology faster than formal schooling, it would be undesirable to regulate private tutoring. Ultimately, public education and private education should compensate each other to spread SW education through collaborative strategies.

In conclusion, we suggest that the major challenges faced in SW education by K–12 schools in South Korea include changing the culture of teaching and learning coding skills, the need for developing teacher competency in both technical and pedagogical aspects, and the tensions between policy imperatives and school practices. It is expected that this chapter presents information about the current status of SW education in South Korea in terms of both opportunities and challenges, and such information can be used to inform countries with a similar trajectory toward cultivating students' CT through the formal curriculum.

NOTE

1. The term "software education" is the English version of the official naming in Korean "소프트웨어 [software] 교육 [education]" of the concerned policy initiative in Korea.

REFERENCES

Angeli, Charoula, Joke Voogt, Andrew Fluck, Mary Webb, Margaret Cox, Joyce Malyn-Smith, and Jason Zagami. 2016. "A K-6 Computational Thinking Curriculum Framework: Implications for Teacher Knowledge." *Educational Technology & Society* 19, no. 3: 47–58.

Bell, Tim, Jason Alexander, Isaac Freeman, and Mick Grimley. 2009. "Computer Science Unplugged: School Students Doing Real Computing without Computers." *The New Zealand Journal of Applied Computing & Information Technology* 13, no. 1: 20–29.

Blackwell, Courtney K., Alexis R. Lauricella, Ellen Wartella, Michael Robb, and Roberta Schomburg. 2013. "Adoption and Use of Technology in Early Education: The Interplay of Extrinsic Barriers and Teacher Attitudes." *Computers & Education* 69: 310–319. https://doi.org/1016/j.compedu.2013.07.024.

Davis, Fred D. 1989. "Perceived Usefulness, Perceived Ease of Use and User Acceptance of Information Technology." *MIS Quarterly* 13, no. 3: 319–339. https://doi.org/10.2307/249008.

Donnelly, Dermot, Oliver McGarr, and John O'Reilly. 2011. "A Framework for Teachers' Integration of ICT into Their Classroom Practice." *Computers & Education* 57, no. 2: 1469–1483. https://doi.org/10.1016/j.compedu.2011.02.014.

Ertmer, Peggy A. 1999. "Addressing First- and Second-Order Barriers to Change: Strategies for Technology Integration." *Educational Technology Research & Development* 47, no. 4: 47–61. https://doi.org/10.1007/BF02299597.

Ertmer, Peggy A. 2005. "Teacher Pedagogical Beliefs: The Final Frontier in Our Quest for Technology Integration?" *Educational Technology Research & Development* 53, no. 4: 25–39. https://doi.org/10.1007/BF02504683.

Ham, Seong-Jin, Soonhwa Kim, and Ki-Sang Song. 2014. "Development of CT-STEAM Education Program Enhancing Integrated Thinking Skills for Elementary School." *The Journal of Korean Association of Computer Education* 17, no. 6: 81–91.

Hansen, Jens Jørgen, and Lea Tilde Rosenlund. 2018. "Teaching in a Networked World–Skills, Knowledge and Beliefs." In *Designing for Learning in a Networked World*, edited by Nina Bonderup Dohn, 81–101. London: Routledge.

Hsu, Ting-Chia, Shao-Chen Chang, and Yu-Ting Hung. 2018. "How to Learn and How to Teach Computational Thinking: Suggestions Based on a Review of the Literature." *Computers & Education* 126: 296–310. https://doi.org/10.1016/j.compedu.2018.07.004.

Hur, Young. 2019. "Development of Unplugged Coding Education Program for the Elementary School." *Korean Society of Basic Design & Art* 20, no. 1: 586–597.

Hwang, Yohan, Mun Kongju, and Park Yunebae. 2016. "Study of Perception on Programming and Computational Thinking and Attitude toward Science Learning of High School Students through Software Inquiry Activity: Focus on Using Scratch and Physical Computing Materials." *Journal of the Korean Association for Science Education* 36, no. 2: 325–335. https://doi.org/10.14697/jkase.2016.36.2.0325.

Jeon, Myounghoon, Maryram Fakhr Hosseini, Jaclyn Barnes, Zackery Duford, Ruimin Zhang, Joseph Ryan, and Eric Vasey. 2016, March. "Making Live Theatre with Multiple Robots as Actors: Bringing Robots to Rural Schools to Promote STEAM Education for Underserved Students." Paper presented at the 11th ACM/IEEE International Conference on Human Robot Interaction. Christchurch, New Zealand. https://doi.org/10.1109/HRI.2016.7451798.

Jung, Min Kyung, and Park Sun Mi. 2018. "The Effects of STEAM Activities Using Unplugged Computing on Young Children's Creativity and Problem Solving Ability." *Journal of Learner-Centered Curriculum & Instruction* 18, no. 3: 705–724.

Kale, Ugur, and Debbie Goh. 2014. "Teaching Style, ICT Experience and Teachers' Attitudes toward Teaching with Web 2.0." *Education & Information Technologies* 19, no. 1: 41–60. https://doi.org/10.1007/s10639-012-9210-3.

Kane, Ruth, Susan Sandretto, and Chris Heath. 2002. "Telling Half the Story: A Critical Review of Research on the Teaching Beliefs and Practices of University Academics." *Review of Educational Research* 72, no. 2: 177–228. https://doi.org/10.3102/00346543072002177.

Koehler, Matthew J., and Punya Mishra. 2009. "What Is Technological Pedagogical Content Knowledge?" *Contemporary Issues in Technology and Teacher Education* 9, no. 1, 60–70.

Kim, Han-sung, and Soo-jin Jun. 2019. "Importance-Performance Analysis of Tasks and Policies of Elementary School Teachers for SW Education." *The Journal of Educational Information & Media* 25, no. 1: 151–170. https://doi.org/10.15833/KAFEIAM.25.1.151.

Kim, Jeongmin, Ilkyung Hong, and Kyungmin Kim. 2016. "A Study on Teaching Materials Development for Computational Thinking through Play." *The Korean Association of Computer Education* 20, no. 2: 187–190.

Kim, Seokhee. 2015. *Development and Effect Study of Creative Maker STEAM Program Based on Physical Computing* [Teacher Community Research Report]. Seoul, Korea: Foundation for the Advancement of Science & Creativity (KOFAC).

Korea Education & Research Information Service. 2017a. "White Paper on ICT Education in Korea, 2017." Accessed October 10, 2020. http://lib.keris.or.kr/bbs/list/6.

Korea Education & Research Information Service. 2017b. *Results of Master Teacher Education in 2017 Software Education*. Daegu, Korea: Korea Education and Research Information Service [KERIS].

Korea Educational Development Institute. 2019. *2019 Basic Education Statistics*. Chungbuk, Korea: Korea Educational Development Institute [KEDI].

Krauss, Jane, and Kiki Prottsman. 2017. *Computational Thinking and Coding for Every Student: The Teacher's Getting-Started Guide*. Thousand Oaks, CA: Corwin Press.

Le Fevre, Deidre M. 2014. "Barriers to Implementing Pedagogical Change: The Role of Teachers' Perceptions of Risk." *Teaching & Teacher Education* 38: 56–64. https://doi.org/10.1016/j.tate.2013.11.007.

Lee, ChangKwon, Jaechoon Jo, and HyeonCheol Kim. 2019. "A Study on Gender Difference of SW Recognition by Middle School Students." *The Journal of Korean Association of Computer Education* 22, no. 1: 11–20. https://doi.org/10.32431/kace.2019.22.1.002.

Lee, Jeongmin, and Eunji Ko. 2018. "The Effect of Software Education on Middle School Students' Computational Thinking." *The Korea Contents Society* 18, no. 12: 238–250. https://doi.org/10.5392/JKCA.2018.18.12.238.

Lee, Jeongmin, Hyunmin Chung, and Eunji Ko. 2018. "Structural Relationships among Factors Affecting Teachers' Robot-Based SW Education Acceptance in Primary School." *The Korea Contents Society* 18, no. 5: 215–229. https://doi.org/10.5392/JKCA.2018.18.05.215.

Lee, Jeongmin, Hyeonkyeong Park, and Hyungshin Choi. 2018. "Effects of SW Education Using Robots on Computational Thinking, Creativity, Academic Interest and Collaborative Skill." *Korean Association of Information Education* 22, no. 1: 9–21. https://doi.org/10.14352/jkaie.2018.22.1.9.

Lee, Soyul, and YoungJun Lee. 2019a. "The Analysis of Difference in Software Education Teaching Efficacy According to Variables of In-Service Elementary School Teachers." *The Journal of Korean Association of Computer Education* 22, no. 4: 1–10. https://doi.org/10.32431/kace.2019.22.4.001.

Lewin, Cathy, and Amina Charania. 2018. "Bridging Formal and Informal Learning through Technology in the Twenty-First Century: Issues and Challenges." In *Second Handbook of Information Technology in Primary and Secondary Education*, edited by Joke Voogt, Gerald Knezek, Rhonda Christensen, and Kwok-Wing Lai, 199–215. Springer, Cham. https://doi.org/10.1007/978-3-319-71054-9_13.

Ministry of Education in South Korea. 2018. "2018 Ministry of Education Busi-ness Manual." Accessed October 10, 2020. http://www.moe.go.kr/boardCnts/view .do?boardID=72703&lev=0&statusYN=C&s=moe&m=0606&opType=N&boardSeq =73294.

Ministry of Education in South Korea. 2015. "Press Release: The Blueprint for Software Education from Elementary School to University (in Korean)." Accessed October 10, 2020. https://www.moe.go.kr/boardCnts/view.do?boardID=339&lev =0&statusYN=C&s=moe&m=02&opType=N&boardSeq=60059.

Ngamkajornwiwat, Potiwat, Pat Pataranutaporn, Werasak Surareungchai, Bank Ngamarunchot, and Tara Suwinyattichaiporn. 2017, December. "Understanding the Role of Arts and Humanities in Social Robotics Design: An Experiment for STEAM Enrichment Program in Thailand." Paper presented at 2017 IEEE 6th International Conference on Teaching, Assessment, and Learning for Engineering (TALE). Hong Kong, SAR, China. http://doi.org/10.1109/TALE.2017.8252378.

Noh, Jiyae. 2018. "The Difference of Computational Thinking and Attitudes toward Robots According to Assistant Teacher in SW Education Using Robot." *Journal of the Korean Association of Information Education* 1, no. 1: 307–316. http://doi.org/10 .14352/jkaie.2018.22.3.307.

Noh, Jiyae, and Jeongmin Lee. 2017. "The Effects of SW Education Using Robot on Computational Thinking." *Korean Association of Information Education* 1, no. 1: 285–296. http://doi.org/10.17232/KSET.34.3.849.

Park, Eunkyung. 2016. "Preparing Students for South Korea's Creative Economy: The Successes and Challenges of Educational Reform." Asia Pacific. Accessed October 10, 2020. https://www.asiapacific.ca/sites/default/files/filefield/south_korea_education _report_updated.pdf.

Park, Hyeongyon, Sunghun Ahn, Chongmin Kim, and Hyunjung Lim. 2017. "Analy-sis of Influencing Factors of Elementary School Students' Computational Thinking and SW Education Attitudes Using 3-Level Multilevel Models." *The Journal of Korean Association of Computer Education* 20, no. 6: 83–94.

Ryoo, Dahyeon, Hyo-Jeong So, and Dongsim Kim. 2019, June. "The Complexity of Teacher Knowledge, Skills and Beliefs about Software Education: Narratives of Korean Teachers." Paper presented at the International Conference on CTE 2019. Hong Kong, SAR, China.

Saeli, Mara, Jacob Perrenet, Wim M.G. Jochems, and Bert Zwaneveld. 2011. "Teach-ing Programming in Secondary School: A Pedagogical Content Knowledge Perspec-tive." *Informatics in Education* 10, no. 1: 73–88.

So, Hyo-Jeong, and Bosung Kim. 2009. "Learning about Problem-Based Learning: Student Teachers Integrating Technology, Pedagogy and Content Knowledge." *Aus-tralasian Journal of Educational Technology* 25, no. 1: 101–116.

So, Hyo-Jeong, Dongsim Kim, and Dahyeon Ryoo. 2020. "Trajectories of Developing Computational Thinking Competencies: Case Portraits of Korean Gifted Girls." *The Asia-Pacific Education Researcher* 29: 85–100. https://doi.org/10.1007/s40299-019 -00459-z.

Sullivan, Amanda, and Marina Umaschi Bers. 2016. "Robotics in the Early Childhood Classroom: Learning Outcomes from an 8-week Robotics Curriculum in Pre-Kindergarten through Second Grade." *International Journal of Technology & Design Education* 26, no. 1: 3–20. https://doi.org/10.1007/s10798-015-9320-5.

Tae, Jinmi. 2016. "Development of STEAM Education Program Utilizing Software Teaching Tool for Elementary School Pupils." *The Journal of the Korean Society for the Gifted & Talented* 15, no. 2: 121–147. https://doi.org/10.17839/JKSGT.2016.15.2.121.

Tedre, Matti, and Peter Denning. 2016, November. "The Long Quest for Computational Thinking." In *Proceedings of the 16th Koli Calling International Conference on Computing Education Research*, edited by Judy Sheard and Calkin Suero Montero, 120–129. New York: ACM.

Thies, Renate, and Jan Vahrenhold. 2013, March. "On Plugging Unplugged into CS Classes." Paper presented at the 44th ACM Technical Symposium on Computer Science Education. Denver, CO. https://doi.org/10.1145/2445196.2445303.

Yoo, Inhwan, YoungKwon Bae, Wooyeol Kim, Jaeho Choi, Maneui Kim, and Jaecheon Jeon. 2018. "The Development and Application of STEAM Education in Robot Programming." *Korean Association of Information Education* 9, no. 1: 157–164.

II

SCHOOL IMPLEMENTATION AND CLASSROOM PRACTICE OF COMPUTATIONAL THINKING CURRICULA IN K–12

9

LEARNING COMPUTATIONAL THINKING IN CO-CREATION PROJECTS WITH MODERN EDUCATIONAL TECHNOLOGY: EXPERIENCES FROM FINLAND

Kati Mäkitalo, Matti Tedre, Jari Laru, Teemu Valtonen, Megumi Iwata, and Jussi Koivisto

INTRODUCTION

Living and working in a digitalized society requires skills and competences that are often referred to as twenty-first-century skills. One of those new skills, computational thinking (CT), has recently become a catchphrase for a broad variety of efforts to bring computing skills and knowledge into the school curriculum. Computational thinking is seen not only as a part of computer science courses but as a part of all school subjects (Guzdial 2015; Tedre and Denning 2016; Mouza et al. 2017; Lockwood and Mooney 2017; Denning and Tedre 2019). Computational thinking promises the skills and competences necessary for understanding, controlling, and automating information processes as well as for interpreting the world as information processes (Denning and Tedre 2019, 4). The introduction of CT to K–12 schools, curriculum guidelines, and frameworks have been prepared by major organizations, including the Computer Science Teachers Association (CSTA) in the US, Computing at School (CAS) in the UK, the Australian Curriculum, Assessment and Reporting Authority (ACARA), as well a large number of other national bodies around the world. The Finnish national basic education core curricula (The National Core Curriculum 2014) now include computing—although not as a separate subject to be taught but as a cross-curricular topic that should be integrated in all subjects.

After a long history in academia and the school system (Denning and Tedre 2019, 175–192), the latest wave of CT for K–12 started in 2006 when Jeannette Wing leveraged her position at the US National Science Foundation (NSF) to campaign her CT vision to the public, funding bodies, and academic audiences (Wing 2006). Wing persuaded decision-makers to intensify CT efforts in schools in many countries, starting with the US. Wing's efforts in NSF directorship led to some impressive achievements, such as a new Advanced Placement (AP) program in the US, billions of dollars invested by the Obama administration in CS for All (NSF 2016), the training of ten thousand computing teachers in the US, and a surge of popular initiatives like code.org, k12cs.org, and Google for Education. The latest literature surveys list hundreds of recent articles, textbooks, and journal special issues (Lockwood and Mooney 2017; García-Peñalvo et al. 2016). We can find several early models and tools for testing CT skills, such as the Mobile CT model (Sherman and Martin 2015), the Progression of Early CT model (Seiter and Foreman 2013), and the Real-Time Evaluation and Assessment of CT tool (Koh et al. 2014). Dagienė's CT-related "Bebras Challenge" has been utilized worldwide by more than two million learners (bebraschallenge.org). The International Computer and Information Literacy Study (ICILS) has included CT in its evaluation since 2018 (Bundsgaard et al. 2019). Despite all this, there is a consensus in the "CT for K–12" community that there are remarkable gaps in the state-of-the-art knowledge about CT in K–12 schools. The research literature has repeatedly highlighted three critical problems.

The first problem, which Denning called one of the "remaining trouble spots" with CT education, is: "How do we measure learners' computational abilities?" (Denning 2017). For example, Mark Guzdial's (2015) quintessential textbook on computing education barely touches the topic of how to evaluate CT skills. A 2017 literature review of more than two hundred studies noted that despite some models for testing CT skills, work for testing CT and the existing models are at their early stages of development (Lockwood and Mooney 2017). One of the reports identified a few pioneering articles, each of which urged researchers to turn their attention to how to measure CT skill development (Mannila et al. 2014). Those problems are due to a lack of consensus over what skills CT consists of and what skill progression in CT looks like.

The second problem is: "How do we integrate CT in K–12 subjects?" Typically, CT is taught in specialized computing courses rather than being integrated in school subjects. While there is less empirical research on whether and how the synergies from integrating CT into K–12 subjects promotes students' analytical skills or widens their understanding about CT (Guzdial 2015, 41–51), integrating CT in subjects is crucial for understanding the ways in which digitalization and computerization have changed all sciences and areas of life. In line with the educational vision of teaching disciplinary ways of thinking and practicing, the knowledge and skills of CT fit well the formal learning goals of other subjects. However, due to fewer concrete examples of how to integrate CT into K–12 subjects, there is a need for empirical evidence on how CT integration influences learners' learning outcomes. Promoting CT in K–12 is challenging due to its specialized nature and broad applicability. Few K–12 teachers have the requisite knowledge and skills of CT as a concept or in its integration in other subjects, and while computer scientists have knowledge and skills in CT in computing, they lack the knowledge and skills for CT integration in education (Mouza et al. 2017).

The third critical problem has to do with lack of understanding: "How do teachers learn to synthesize CT with existing content and pedagogical strategies?" (cf. Mouza et al. 2017). We need to understand the technological knowledge and skills needed for teachers to integrate technology and technological resources effectively in classrooms. We have examples of studies on how teachers combine the areas of technology and pedagogy with the content taught (see Koehler and Mishra 2009). Valtonen et al. (2019) show that despite intensive integration of technology, content, and pedagogy, pre-service teachers experience technological knowledge as a separate domain. Studies show that integrating CT in teacher education courses enhances pre-service teachers' knowledge and understanding of CT, but when CT is taught separately from teachers' own disciplines, that understanding remains at an abstract level and is not applied in teaching (Yadav, Stephenson, and Hong 2017). We can create a framework that offers a practical model for integrating CT within those subject matters and pedagogical approaches that teachers are expected to teach in classrooms (Yadav et al. 2017). However, CT literature has pointed out a dire need for more research on understanding teacher learning and the best ways to support it (Guzdial 2015; Mouza et al. 2017).

This chapter presents two example cases to broaden our understanding of how learning CT ideas can be supported through technology and classroom practices in the context of STEAM topics and what leeway CT gives to STEAM classes. It also presents what one can learn from nonformal education in an after-school programming club. We investigate learners' as well teachers' CT knowledge and skills through the cases provided here.

CASE 1: PROMOTING COMPUTATIONAL THINKING WITH MINECRAFT IN A K–12 PROGRAMMING CLUB

This case study was designed to integrate informal learning activities and computational thinking concepts for learners in the context of an after-school Minecraft club. The design rationale for the Earth 2.0 Minecraft game was recurring difficulties in finding a functional and motivating way to teach computational thinking and programming in an after-school club (Koivisto, Laru, and Mäkitalo 2019).

To solve this issue, a pedagogically and theoretically grounded Minecraft game learning experience was designed and evaluated with the club participants in an authentic setting. Minecraft is a multiplayer sandbox game designed around breaking and placing blocks. Unlike many other games, when played in its traditional settings, Minecraft grants players the freedom to immerse themselves into their own narrative: to build, create, and explore. Minecraft, along with modification software ("mods"), has the tools for teaching and learning programming (Zorn et al. 2013; Risberg 2015; Nebel, Schneider, and Rey 2016; Näykki et. al 2019).

TOOLS

The tools used were the Minecraft game, Earth 2.0 map, and seven Minecraft modifications (mods). Earth 2.0 includes problem-based puzzles embedded in an engaging post-apocalyptic narrative. Modifications enable one to modify Minecraft's eighteen game rules, alter game content, redesign textures, and give players new abilities (Kuhn and Dikkers 2015). While Earth 2.0 provided a context for the game narrative and gameplay itself, modifications worked as an engine to enable interactivity and programming during the game.

The most important modification in this context was Computer-CraftEdu. It is an extension based on Dan200's ComputerCraft but with some added features, including the programmable Beginners Turtle (see table 9.1), which learners are able to program with descriptive icons. Otherwise, learners must know the text-based LUA language, which is used in both extensions. According to Wilkinson, Williams, and Armstrong (2013), the use of ComputerCraft has a positive impact on learning and motivation in programming. All mods used in this experiment are briefly explained in table 9.1.

PARTICIPANTS

Participant groups enrolled in after-school programming clubs for three months with the first four sessions on Minecraft. The data were collected from five clubs, each consisting of eight to twenty participants between seven and twelve years of age, with altogether sixty-two participants (nine female and fifty-three male). All five groups used the prototype version of the Earth 2.0 computational thinking game[1] (Koivisto, Laru, and Mäkitalo 2019).

Table 9.1 All the mods used to expand original Minecraft gameplay

Modification	Purpose of the modification in Earth 2.0
Forge	Minecraft modding API
ComputerCraft	Virtual computers for programming scripted events
ComputerCraftEdu	Educational version of ComputerCraft that includes the Beginners Turtle.
CustomNPCs	Characters to make interaction between player and the game world possible and meaningful
Malisisdoors and Malisiscore	Code-locked and animated doors
Optifine	Minecraft's optimization to improve performance
JammerCraft	More realistic and detailed textures that match with the narrative of Earth 2.0

TASKS AND PEDAGOGICAL DESIGN

In the Earth 2.0 Minecraft world, the game narrative starts from a problem the players face. Players should program robots to reconstruct the post-apocalyptic Earth, but suddenly all activities stop. Players are assigned the role of scientist-astronaut, and they are instructed to go back to the Earth and investigate the situation. On Earth, they find information about a person who is trying to sabotage the scientist-astronauts' attempts to rehabilitate the planet Earth.

The game narrative is further divided into four main puzzles (quests) and one bonus puzzle (quest), which all are separate Minecraft worlds and one bonus world. First, players must take the tutorial to learn to use the programming tool, Beginners Turtle. Then they are introduced to basic programming concepts and practices in a specific order (see table 9.2). Concepts and practices of computational thinking are distributed to three sequential task groups: Quests 1, 2, and 3. Finally, a fourth task group, Bonus Quest, is included for more advanced players (see detailed descriptions in figure 9.1 and table 9.2). The layout of an individual puzzle is presented in figures 9.2 and 9.3.

The Earth 2.0 experiment in the context of the programming club is an example of using off-shelf sandbox games as a tool to teach CT concepts and practices for inexperienced learners in the K–12 context. According to Lye and Koh (2014), it is dubious to assume that learners could figure out computational practices and perspectives through pure self-discovery. Support from more experienced learners and teachers helps with understanding better computational practices. So, it is necessary to bring all participants on the same level of the basic ideas of programming and computational thinking by using different scaffolds that restrict original gameplay, although it violates the core principles and ideas of the constructivist gameplay (Kafai and Burke 2015; Lye and Koh 2014; Mayer 2015).

CASE 2: K–12 TEACHERS' AND FAB LAB FACILITATORS' ATTITUDES AND BELIEFS ABOUT COMPUTATIONAL THINKING IN DIGITAL FABRICATION AT FAB LAB

In this case study, the aim was to explore learners' digital fabrication activities at Fab Lab Oulu, located in the University of Oulu, Finland. The

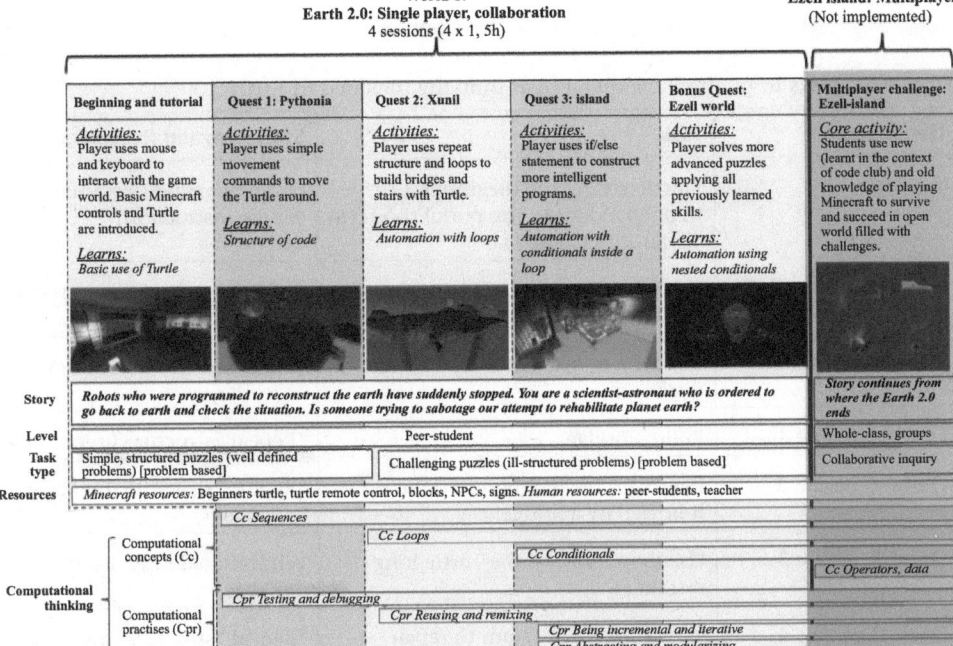

9.1 The pedagogical design, where computational concepts and practices were integrated as a part of the Minecraft gameplay.

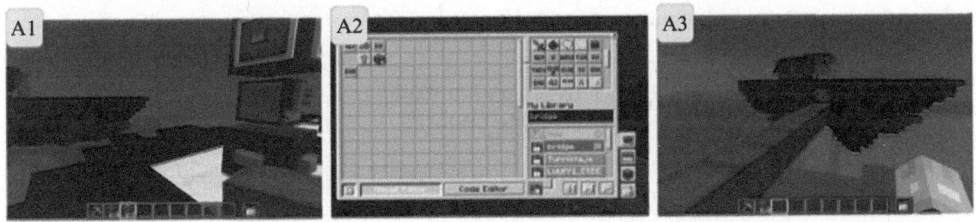

9.2 Quest 2, puzzle 1. A1: Player encounters a problem. A2: Player programs the Turtle to build a bridge. A3: Player executes the program, and Turtle builds a bridge for the player.

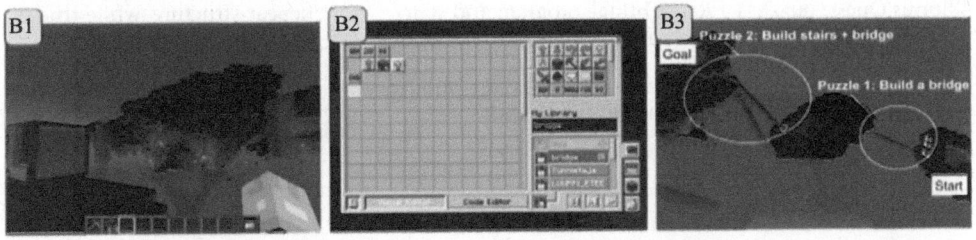

9.3 Quest 2, puzzle 2. B1: Player encounters a slightly different problem. B2: Player remixes the bridge program to build stairs. B3: Overview of puzzles 1–2 in Quest 2, both puzzles completed.

Table 9.2 Tasks to promote computational thinking included in Earth 2.0

Game phase	Task	Structures and statements used
Tutorial	Practice using the (Beginners) Turtle to open a door to the portal room (to advance to Quest 1).	Only movement command (single instruction)
Quest 1: puzzles 1–3	Use movement commands to move the Turtle and advance.	Only movement commands (sequential execution)
Quest 1: puzzle 4	Use loop to move the Turtle long distances.	While-true-do (iteration)
Quest 2, puzzle 1	Build a bridge.	"Repeat" structure (iteration)
Quest 2, puzzle 2	Remix bridge program to build stairs followed by a bridge.	"Repeat" structure (iteration)
Quest 2, puzzle 3	Use loop to move the Turtle long distances.	While-true-do (iteration)
Quest 2, puzzle 4	Remix bridge program to repair broken bridge.	"Repeat" structure (iteration)
Quest 3, puzzle 1	Practice to use conditionals and loops together to dig through a cave.	While-true-do, (iteration) If-then-else (conditional branching / selection)
Quest 3, puzzle 2	Program automated Turtle to avoid obstacles.	While-true-do, (iteration), if-then-else (selection)
Quest 3, puzzle 3	Remix previous program to build a bridge while avoiding obstacles.	While-true-do, (iteration), if-then-else (selection)
Quest 3, puzzle 4	Program automated Turtle to solve a labyrinth.	While-true-do, if-then-else if
Quest 3, puzzle 5	Reuse and debug previous program to solve bigger labyrinth.	While-true-do, if-then-else if
Quest 3, puzzle 6	Remix bridge, obstacle, and digger programs to reach the exit.	While-true-do, if-then-else if
Bonus Quest, puzzle 1	Reuse bridge program and stair program to create path to the exit.	Repeat-structure, while-true-do
Bonus Quest, puzzle 2	Create intelligent bridge-builder program to create path.	While-true-do, if-then-else if

Fab Lab offers digital fabrication facilities for K–12 schools in the Oulu area. Those activities typically combine 2D and 3D designing and manufacturing, prototyping with electronics, basic embedded systems programming, and using various manufacturing tools and machines. Digital fabrication activities combine content and skills from multiple subjects, and the activities are typically ill structured and student centered (Iwata et al. 2020; Näykki et al. 2019; Pitkänen, Iwata, and Laru 2020).

In this case study, seventh- to ninth-grade learners from three K–12 schools in Oulu participated in digital fabrication activities that spanned from three to five days at Fab Lab Oulu (see table 9.3). Learners worked on digital fabrication projects in groups of two to five. Activities were led by the Fab Lab instructors and were observed by teachers. The activities were

Table 9.3 Participants and project details of three case schools

Participants Project details	Case I: School A	Case II: School B	Case III: School C
Number of participants	12 (15–16 years), 1 teacher	20 (13–15 years), 2 teachers	9 (15–16 years), 2 teachers
Duration	5 days	3 days	5 days
Projects	Useless box, Rail for camera, Electronic controlled lock, Jukebox game, Music car	Finland's 100-year anniversary calendar, Finland 100-years history wheel, Finnish flag day clock	Two models of playhouse
Required conditions	Use Arduino Uno board and at least one actuator, fabricate mechanics using laser cutter or 3D printer, make functional artefact	Use Arduino Uno board, fabricate mechanics using laser cutter	A competition between two teams designing a playhouse for the school community
Software for designing	Inkscape, Autodesk TinkerCad	Inkscape	Inkscape, SketchUp
Machines	Laser cutter, 3D printer	Laser cutter	Laser cutter, vinyl cutter, sewing machine
Electronics	Arduino Uno board, servos, buttons, piezoelectric buzzer	Arduino Uno board, servos	–
Programming	Arduino Software (IDE)	Arduino Software (IDE)	–

loosely structured with minimal instruction, and the activity designs differed among schools. The instructors gave assistance only when learners had problems in the process. Two focus group interviews were conducted after the digital fabrication activities. Three teachers from two schools (focus group interview I) and two instructors at the Fab Lab (focus group interview II) participated in the focus group interviews.

In the focus group interview, teachers described the ways digital fabrication activities make an effective context for integrating CT with knowledge and skills from multiple subjects. One of the teachers revealed the following: "this kind of working [project-based digital fabrication activities] brings together very many school subjects. There are mathematics, physics, coding, programming." That teacher specifically linked programming with the learning goals of the national core curriculum for basic education.

Teachers explained the benefits of tangible project-based activities in relation to learning CT: Learning CT becomes more effective when learners have concrete projects where they can apply knowledge into practice. When learner groups had challenges using the microcontroller, the instructor gave short lectures on how the logical ports of the microcontroller work. One teacher described that those lectures helped learners develop an understanding of how computers work, which they can utilize in their projects on the fly. In project-based activities, learners were able to test their CT knowledge in real projects.

Teachers found that the digital fabrication activities that combine hardware and software into one project are effective for learning CT. They felt that the processes of making physical objects enhanced learners' logical thinking (their projects made use of logic gates). Although learners have a general and abstract idea of the processes of making a physical object, they often confront many practical challenges during digital fabrication activities. When dealing with physical parts, learners need to think in concrete, sequential, and interconnected steps to complete their digital fabrication projects. Teachers reflected that "when we started, there was already the idea that it's going to be a process of several steps . . . they [the learners] just didn't know what's going to happen in between those steps." This teacher stated that previously he thought combining hardware and programming is challenging, and therefore he gave ready-made parts that learners only needed to assemble. After visiting the Fab Lab, he shifted his perspective, explaining that in the future learners have the responsibility to plan, design, and create the hardware parts by themselves. In the Fab Lab, teachers observed learners' activities and their performance, and, based on the observations, he decided to change his teaching practice.

However, the same case study also exposed challenges for learning CT in digital fabrication activities (see table 9.4). Firstly, the teachers' discussion in the focus group regarding CT was still at the surface level. During the focus group interview, teachers mentioned several aspects of

Table 9.4   CT practices identified in focus group interviews

	Focus group interview I (N=8,387)		Focus group interview II (N=6,328)	
CT practices(Barr, Harrison, and Conery 2011)	%	N	%	n
1) Formulating problems in a way that computers and other tools can help solve them	45.8%	432	17.8%	147
2) Logically organizing and analyzing data	18.1%	171	28.9%	239
3) Representing data through abstractions	0.0%	–	0.0%	–
4) Automating solutions through algorithmic thinking	7.1%	67	17.8%	147
5) Identifying, analyzing, and implementing possible solutions with the most efficient and effective combination	29.0%	274	35.5%	293
6) Generalizing and transferring this problem-solving process	0.0%	–	0.0%	–
Total	100%	944	100%	826

Note: N shows the total number of words in the focus group interview; n shows the number of words at the node.

CT. These aspects can be categorized based on Barr et al.'s work (2011, 21), which are (1) "formulating problems in a way that computer and other tools can help solve them," (2) "identifying, analyzing, and implementing possible solutions with the most efficient and effective combination," and (3) "logically organizing and analyzing data." Nevertheless, none of those CT aspects were discussed more deeply in the focus group interviews. The instructors who facilitated the digital fabrication activities mentioned that K–12 teachers are unfamiliar with computational concepts and practices as they are not embedded in the curriculum and school culture (Iwata et al. 2020).

Secondly, teachers indicate needs for pedagogy in ill-structured digital fabrication activities to enhance learning (Pitkänen, Iwata, and Laru 2020). Currently, K–12 teachers are not actively involved in the preparation and implementation processes of activities at the Fab Lab, which require technical knowledge and skills beyond teachers' competences. Low involvement might be due to Fab Lab activities, which are not embedded in the curriculum. This case study illustrates that there is a need to involve teachers more in these kinds of activities to enable them to integrate CT with pedagogical approaches as well as assessing CT based on the goals of the curriculum.

DISCUSSION

Our case studies are a part of a larger attempt at integrating CT into Finnish education. They are also aimed at broadening the current understanding of twenty-first-century teacher education by producing empirical results on the effect of Fab Lab interventions on teachers' CT knowledge and skills and using games as a learning environment for CT skills.

The Minecraft example (first case study) demonstrates how CT ideas can be embedded in a constructionist computer game. However, if concepts and practices of CT are not explicit, visible, and measurable, learning CT remains at the surface level for both learners and teachers. Minecraft is an immersive 3D game, which helps teachers and learners to design experiences in which concepts of CT can be very visible and explicit. In that first case study, to complete each quest (challenges), the learner's task was to program Beginner's Turtle to, for example, build a bridge.

During this activity, learners learn both the concepts (e.g., repeat structures) and the practices (e.g., remixing and debugging). Microsoft has adopted a similar approach in its Minecraft Education Edition "Hour-of-Code" tutorial,[2] which is oriented to beginners like the world in the first case study was. However, Minecraft would also be a good platform to teach a more holistic approach to CT than these very structured and scaffolded tutorials oriented to teaching coding skills. Next, we should consider what kind of support mechanisms for helping teachers and students could be designed for adopting a holistic approach to CT concepts, skills, and practices in Minecraft.

The second case study explored K–12 teachers' and Fab Lab facilitators' attitudes and beliefs about CT in the context of digital fabrication. Ill-structured digital fabrication activity, which involves multiple subject matters, is complex context, but it provides a good opportunity for the integration of CT in the school curriculum. Project-based activities where learners deal with hardware parts, components, and software, require, for instance, decomposing complex problems into subproblems and logically organizing solutions into steps to achieve the goals.

However, this second case study indicates the need for further developing teachers' CT understanding, especially in terms of integrating CT in the curriculum with appropriate pedagogical approaches. It was found that teachers and facilitators are not fully aware of the concepts of CT. Teachers' understanding covered only the surface of CT, such as using computers to solve problems, while CT practices, which involve the fundamental concepts of CT, such as abstractions and automation, were not intensively discussed (Iwata et al. 2020).

On the other hand, facilitators are already adept at CT skills, and they may not have the awareness of defining CT to be a competence that learners need to develop. Teachers' and facilitators' awareness of the concepts of CT may be essential to providing opportunities for learners to understand and apply CT practices in digital fabrication activities. Insufficient understanding of CT concepts and practices might diminish the potential to develop CT in digital fabrication activities (Iwata et al. 2020).

The findings are in line with Mannila et al. (2014, 14), illustrating that very few teachers implement the concepts of abstraction and automation in classroom practice. Furthermore, other studies have shown that K–12 teachers' understanding about CT competence and skills often remains at a superficial level, typically at the level of basic programming skills (Mouza et al. 2017). Yet, that result is unsurprising in the absence of consensus over even the most fundamental principles of CT in curricular integration (Denning and Tedre 2019). Increased scaffolding efforts turned out to benefit learning, especially before engaging in complex ill-structured problem-solving activities in the context of STEAM projects or in digital fabrication and making (Fab Lab). The question is: Do we need better pedagogical design and increased scaffolding when we try to introduce CT concepts, practices, and perspectives to K–12 education? Do we have adequate pre- and in-service teacher education that would cover these emerging topics?

From the Finnish perspective, the ideas of CT are embedded in our national core curriculum for basic education without naming CT. For instance, key content areas related to the objectives of mathematics in grades 7–9 include: (1) thinking skills and methods, such as logical thinking and discovering rules and dependencies, (2) examining and applying functions, and (3) data processing, such as collecting, structuring, and analyzing data (Finnish National Board of Education 2016, 402–405). Connecting those mentioned skills with problem-solving, which applies how computers work, enhance teachers' understanding of holistic views of CT (Denning and Tedre 2019, xi).

However, a few elements need to come together for better curricular integration of CT in Finland and other countries. First, we need to build a common understanding about the fundamental computational principles necessary for education. Second, we need working models, templates, and exemplars for integrating CT in curricula. Third, we need pedagogical models that make CT competences and skills more visible. Fourth, we need principles for assessing CT within subjects as well as based on the goals of the basic education curriculum. Instead of a technology-driven or programming-driven approach for integrating CT in education, we need a human- and design-driven as well discipline-driven approach to understanding CT in a broader way.

ACKNOWLEDGMENTS

Thank you to the Fab Lab facilitators, participating teachers and learners, and participants and facilitators of the after-school programming clubs.

NOTES

1. Description of Earth 2.0 Minecraft game: https://sites.google.com/oulu.fi/earth20/english.

2. https://education.minecraft.net/hour-of-code.

REFERENCES

Barr, David, John Harrison, and Leslie Conery. 2011. "Computational Thinking: A Digital Age Skill for Everyone." *Learning & Leading with Technology* 38, no. 6: 20–23.

Bundsgaard, Jeppe, Sofie Gry Bindslev, Elisa Nadire Caeli, Morten Pettersson, and Anna Rusmann. 2019. *Danske elevers teknologiforståelse: Resultater fra ICILS-undersøgelsen 2018*. Aarhus, Denmark: Aarhus Universitetsforlag.

Denning, Peter J. 2017. "Remaining Trouble Spots with Computational Thinking." *Communications of the ACM* 60, no. 6: 33–39.

Denning, Peter J., and Matti Tedre. 2019. *Computational Thinking*. Cambridge, MA: MIT Press.

Finnish National Board of Education. 2016. *National Core Curriculum for Basic Education 2014*. Helsinki: Finnish National Board of Education.

García Peñalvo, Francisco José, Daniela Reimann, Maire Tuul, Alyson Rees, and Ilkka Jormanainen. 2016. "An Overview of the Most Relevant Literature on Coding and Computational Thinking with Emphasis on the Relevant Issues for Teachers." Technical Report, TACCLE3 Consortium, Belgium.

Guzdial, Mark. 2015. "Learner-Centered Design of Computing Education: Research on Computing for Everyone." *Synthesis Lectures on Human-Centered Informatics* 8, no. 6: 1–165. San Rafael, CA: Morgan & Claypool.

Iwata, Megumi, Kati Pitkänen, Jari Laru, and Kati Mäkitalo. 2020. "Exploring Potentials and Challenges to Develop Twenty-First Century Skills and Computational Thinking in K-12 Maker Education." *Frontiers in Education* 5: 87. https://doi.org/10.3389/feduc.2020.00087.

Kafai, Yasmin B., and Quinn Burke. 2015. "Constructionist Gaming: Understanding the Benefits of Making Games for Learning." *Educational Psychologist* 50, no. 4: 313–334.

Koehler, Matthew, and Punya Mishra. 2009. "What Is Technological Pedagogical Content Knowledge (TPACK)?" *Contemporary Issues in Technology and Teacher Education* 9, no. 1: 60–70.

Koh, Kyu Han, Ashok Basawapatna, Hilarie Nickerson, and Alexander Repenning. 2014. "Real Time Assessment of Computational Thinking." In *2014 IEEE Symposium on Visual Languages and Human-Centric Computing (VL/HCC)*: 49–52. IEEE.

Koivisto, Jussi, Jari Laru, and Kati Mäkitalo. 2019. "Promoting Computational Thinking Skills in the Context of Programming Club for K-12 Pupils with the Engaging Game Adventure in Minecraft." In *Proceedings of International Conference on Computational Thinking Education 2019*, 48. Hong Kong: Education University of Hong Kong.

Kuhn, Jeff, and Seann Dikkers. 2015. "How Can Third Party Tools Be Used?" In *Teachercraft: How Teachers Learn to Use Minecraft in Their Classrooms*, 123–138. Carnegie Mellon University Pittsburgh: ETC Press.

Lockwood, James, and Aidan Mooney. 2017. "Computational Thinking in Education: Where Does It Fit? A Systematic Literary Review." *arXiv preprint arXiv:1703.07659*. Technical report, National University of Ireland Maynooth.

Lye, Sze Yee, and Joyce Hwee Ling Koh. 2014. "Review on Teaching and Learning of Computational Thinking through Programming: What Is Next for K-12?" *Computers in Human Behavior* 41: 51–61. https://doi.org/10.1016/j.chb.2014.09.012.

Mannila, Linda, Valentina Dagiene, Barbara Demo, Natasa Grgurina, Claudio Mirolo, Lennart Rolandsson, and Amber Settle. 2014. "Computational Thinking in K-9 Education." In *Proceedings of the Working Group Reports of the 2014 on Innovation & Technology in Computer Science Education Conference*: 1–29.

Mayer, Richard E. 2015. "On the Need for Research Evidence to Guide the Design of Computer Games for Learning." *Educational Psychologist* 50, no. 4: 349–353. https://doi.org/10.1080/00461520.2015.1133307.

Mouza, Chrystalla, Hui Yang, Yi-Cheng Pan, Sule Yilmaz Ozden, and Lori Pollock. 2017. "Resetting Educational Technology Coursework for Pre-Service Teachers: A Computational Thinking Approach to the Development of Technological Pedagogical Content Knowledge (TPACK)." *Australasian Journal of Educational Technology* 33, no. 3: 61–76.

Näykki, Piia, Jari Laru, Essi Vuopala, Pirkko Siklander, and Sanna Järvelä. 2019. "Affective Learning in Digital Education—Case Studies of Social Networking Systems, Games for Learning and Digital Fabrication." *Frontiers in Education* 4 (November): 128. https://doi.org/10.3389/feduc.2019.00128.

Nebel, Steve, Sascha Schneider, and Günter Daniel Rey. 2016. "Mining Learning and Crafting Scientific Experiments: A Literature Review on the Use of Minecraft in Education and Research." *Journal of Educational Technology & Society* 19, no. 2: 355–366.

NSF [National Science Foundation]. 2016. "CS for All." Accessed November 23, 2020. https://www.nsf.gov/news/special_reports/csed/csforall.jsp.

Pitkänen, Kati, Megumi Iwata, and Jari Laru. 2020. "Exploring Technology-Oriented Fab Lab Facilitators' Role as Educators in K-12 Education: Focus on Scaffolding

Novice Students' Learning in Digital Fabrication Activities." *International Journal of Child-Computer Interaction* 26. https://doi.org/10.1016/j.ijcci.2020.100207.

Risberg, Cathy. 2015. "More than Just a Video Game: Tips for Using Minecraft to Personalize the Curriculum and Promote Creativity, Collaboration, and Problem Solving." *Illinois Association for Gifted Children Journal*, 44–48. Accessed November 26, 2020. https://studylib.net/doc/14185247/iagc-journal-focus--creativity--critical-thinking-.

Seiter, Linda, and Brendan Foreman. 2013. "Modeling the Learning Progressions of Computational Thinking of Primary Grade Students." In *Proceedings of the Ninth Annual International ACM Conference on International Computing Education Research*, ICER '13, 59–66. ACM.

Sherman, Mark, and Fred Martin. 2015. "The Assessment of Mobile Computational Thinking." *Journal of Computing Sciences in Colleges* 30, no. 6: 53–59.

Tedre, Matti, and Peter J. Denning. 2016. "The Long Quest for Computational Thinking." In *Proceedings of the 16th Koli Calling International Conference on Computing Education Research*, Koli Calling '16, 120–129. ACM.

The National Core Curriculum. 2014. "The Finnish National Board of Education." Accessed September 9, 2018. https://www.oph.fi/fi/koulutus-ja-tutkinnot /perusopetuksen-opetussuunnitelman-perusteet.

Valtonen, Teemu, Erkko Sointu, Jari Kukkonen, Kati Mäkitalo, Nhi Hoang, Päivi Häkkinen, Sanna Järvelä et al. 2019. "Examining Pre-Service Teachers' Technological Pedagogical Content Knowledge as Evolving Knowledge Domains: A Longitudinal Approach." *Journal of Computer Assisted Learning* 35, no. 4: 491–502.

Wilkinson, Brett, Neville Williams, and Patrick Armstrong. 2013. "Improving Student Understanding, Application and Synthesis of Computer Programming Concepts with Minecraft." In *The European Conference on Technology in the Classroom*.

Wing, Jeannette M. 2006. "Computational Thinking." *Communications of the ACM* 49, no. 3: 33–35.

Yadav, Aman, Chris Stephenson, and Hai Hong. 2017. "Computational Thinking for Teacher Education" *Communications of the ACM* 60, no. 4: 55–62.

Zorn, Christopher, Chadwick A. Wingrave, Emiko Charbonneau, and Joseph J. LaViola, Jr. 2013. "Exploring Minecraft as a Conduit for Increasing Interest in Programming." In *Proceedings of the 8th International Conference on the Foundations of Digital Games (FDG 2013)*, 352–359. Chania, Crete, Greece. http://www.fdg2013.org /program/papers/paper46_zorn_etal.pdf.

10

INTEGRATING DESIGN THINKING INTO K–12 COMPUTATIONAL THINKING CLASSROOMS IN TAIWAN: PRACTICES OF COLLABORATIVE ROBOTIC PROJECTS

Ju-Ling Shih

INTRODUCTION

Computational thinking (CT) is generally regarded as problem-solving skills with logical thinking ability and programming techniques. It is mostly treated as an integral part of interdisciplinary curricula instead of a stand-alone subject and is therefore often associated with STEM- and project-based activities in education.

This chapter begins with a broad overview of the existing standards and curriculum efforts in computational thinking education in the Taiwanese K–12 sector and the initiation of innovative instructions in the classrooms.

To illustrate how instructional innovation is formed, a model of the synergy of design thinking and computational thinking is introduced. The two thinking models may be paralleled, twisted, or aligned, and that should be considered with the instructional needs. This is an iterative process working between divergent and convergent thinking, taking the participants through the "discover," "define," "develop," and "deliver" stages. In this process, students solve problems and use programming to tackle situational challenges.

An instructional example of classroom implementation is given, show-ing how the teacher guided the students through the design thinking and computational thinking process, working in groups to carry out an interdisciplinary collaborative drawing board project in training their problem-solving skills. The primary results of the classroom implementation are reported along with the revelation of the possible subsequent instructional challenges.

At the end of the chapter, the teacher training process—"to see, to feel, to change"—is presented to illustrate how the teachers shall be trained from knowing to doing, from now to new, with insight and goals.

K–12 CT EDUCATION IN TAIWAN

In this section, the Darmstadt model (Hubwieser 2013) is used to describe Taiwan's CT education in terms of educational system, curriculum, knowledge, teacher qualification, and teaching methods.

EDUCATIONAL SYSTEM

In Taiwan, students attend elementary school at age six, going through grades 1 to 6 in primary school, grades 7 to 9 in junior high school, and grades 10 to 12 in senior high school, to complete the twelve-year compulsory education. Taiwan's Ministry of Education (MOE) initiated a twelve-year curriculum for basic education in 2014 with the theme of "Spontaneity, Interaction, Common Good" (Tsai et al. 2011) and published the detailed guidelines for the 2018 academic year. The three themes encourage students (1) to do self-learning, think systematically to solve problems, and respond to changing situations for spontaneity; (2) to communicate with others and be aware of the world for interaction; and (3) to care about the health of society and people around them for the common good.

CURRICULUM

The rapid current of information technology development and innovative educational trends have driven the Ministry of Education in Taiwan to continue with educational reform, and it finally categorized "systematic thinking and problem solving" into the autonomous learning realm in the curriculum in 2014 (Ministry of Education 2014). Followed by several iterations of the guidelines and contents, the 2016 version of the curriculum guidelines has added the technology learning area that includes two curriculum tracks: "information technology" and "life and technology."

The "information technology" curriculum is based on traditional computer courses that teach students the use of software such as word processing, slide presentations, graphical presentations, and so on. It uses computational thinking as its core, which nurtures abilities such as "computational thinking and problem-solving," "information technology and co-creation," "information technology and communication," and "attitude toward the use of information technology." The learning content includes six aspects: "algorithms," "coding," "system platforms," "information representation, management, and analysis," "information technology application," and "information technology and human society" (National Academy for Educational Research 2015). The main goal of CT is not to train students to become programmers but to promote logical thinking through the structure of programming language, intuitive cognition of machine processing, the ability to solve problems in a procedural way (Wing 2008), and the ability to explore wider learning approaches (Resnick 2013). Programming guides students to dissect a big problem into small problems to more clearly, accurately, and demonstratively explain the process of solving problems (Fernaeus, Kindborg, and Scholz, 2006 and Scholz 2006; Angeli and Giannakos 2020). These skills can be used in interdisciplinary learning and applied to real-world situations.

On the other hand, the "life and technology" curriculum offers students general technological knowledge, tools, and skills for the purpose of computational thinking, while at the same time enhancing students' cross-disciplinary knowledge integration abilities with maker practices (Ministry of Education 2016). It is a STEM-based curriculum that provides students with the ability to solve real-life complex problems in the realm of science, technology, engineering, and mathematics and trains them to have twenty-first-century key competencies (Kucuk and Sisman 2020).

KNOWLEDGE

In short, the computational thinking curricula are divided into two parts— (1) programming, which involves computers, and (2) making, which emphasizes on hands-on practices—that are practiced in the aforementioned two curricula, respectively. Both of them are mostly done in the cross-disciplinary approach that integrates theme-based learning contents, but neither has predetermined teaching materials or a recommended course structure. From personal experience working with dozens of elementary and secondary schools, all teachers responsible for these curricula are searching for materials that are appropriate to the students they teach and are designing materials specifically for their own school environment and educational goals. For example, some elementary schools are teaching Scratch from the third grade up and teaching App Inventor for advanced levels; some schools emphasize STEM-based or IoT (Internet of Things) applications. The STEM education in Taiwan, defined as cross-disciplines of science, technology, engineering, and mathematics, is mostly done with robots and microcontrollers such as mBot, Lego EV3, Arduino, and Micro:bit. Since these technological tools require programming to make them work, the STEM curriculum is often cross-related to the programming curriculum. In most instances, these are inseparable. Thus, CT is regarded more as a problem-solving process with logistical thinking and done through programming. Psycharis (2013) mentions that the teaching of STEM requires students to explore and process data using computer language, so computing sits in a central position of STEM (Chi and Jain 2011; Henderson, Cortina, and Wing 2007).

In addition to the somewhat blurry line between the two curricula, some schools integrate them into their own school-based curriculum. If the school prioritizes teaching about the nearby temple, community fishery, or forest guardian, then the CT-related skills are integrated into that curriculum, bringing the skills into school-based real-life scenarios. Therefore, CT in Taiwan is no longer a stand-alone subject but is practiced in various forms with one or many teachers' efforts. In some schools, CT is extended into after-school clubs, which often train students to be high performers who participate in all kinds of national and international competitions. The whole practice matches the global educational spirit stated by Taiwan's MOE: cross-field integration, real-life scenarios, school-based curriculum, and personalized learning.

TEACHER QUALIFICATION

These courses are taught by qualified teachers with or without technology or information professional training due to uneven teacher allocations in rural and urban schools. Since the needs in different schools vary, teachers form their own groups to reach updated information and self-train to teach. They either join groups formed through the educational bureau in every city, which appoints head teachers to take the lead, or individual teachers initiate interest groups, study groups, work groups, exchange groups, or workshops and conferences for the whole country. Therefore, the teachings are of various innovative ways and are mostly bottom-up.

TEACHING METHODS

Nevertheless, Frymier, Shulman, and Houser (1996) also observe in the classrooms that teachers usually control the class content while students are simply content followers. Similar situations can be seen in the STEM and robotics education in Taiwan where students follow the manuals provided by the teachers, digital or paper, to assemble robots and code for actions to perform uniform tasks. The completion of tasks is deemed "learning" regardless of its implications.

General problems perceived in today's classroom practices include:

1. In the classroom, teachers normally provide self-made teaching materials and manuals to the students to follow the instructions and imitate the sample projects. Learning evaluation is based on the completion rate of the projects and sometimes the speed of completion.
2. The students often complete their projects knowing how without knowing why.
3. The students have few opportunities to be creative and make their own creations.
4. The project scenario is predesigned by the teachers, real or fictional, with either tasks or problems. Sometimes the scenario is dismissed in the teaching, which leaves only the robotic tasks to be done.

Donovan and Bransford (2005) describe the manuals as recipes that give lock steps to the process, which shortchanges the imagination. Questions, problems, and situations should be more unclear so that students are required to search for clues and innovative solutions. Thus, this chapter proposes a pedagogical framework to provide another type of instructional model that would address the current needs of CT education.

INTEGRATE COMPUTATIONAL THINKING WITH DESIGN THINKING

Computational thinking was defined as using computer logic to solve problems. It draws on the concepts fundamental to computer science and outlines five cornerstones of computational thinking (Selby and Woollard 2013; Wing 2006).

(CT 1) decomposition: breaking a big problem down into smaller parts.
(CT 2) pattern recognition: looking for similarities within and between problems.
(CT 3) abstraction: taking the details out of a problem and ignoring irrelevant information.
(CT 4) algorithms: create the simple step-by-step rules to follow in order to solve the problem.
(CT 5) generalization: adapting solutions to current problems to solve new ones.

However, CT is generally recognized in two ways: specific CT refers to computer programming skills, while generic CT refers to overall problem-solving skills. But Wing (2006) more precisely said that CT involves solving problems, designing systems, and understanding human behavior, so it is necessary to perceive CT as computer programming, scenario connection, and interdisciplinary learning, which is a comprehensive view of both specific CT and generic CT.

DESIGN THINKING AS PROBLEM-SOLVING PROCESS

While CT describes the techniques of each production or problem-solving stage, design thinking (DT) guides one through the production process, providing insight to approach the problem. The major step of DT that is distinct from other teaching models is to start from the "empathy" stage before diagnosing the problems and ideating for solutions. For many classroom practices, students deal with targeted projects without having to think about how and why the projects would be made in a certain way.

Design thinking is a thinking model that places creating a humanistic environment as top priority and treats convenience and problem-solving with creativity as the key goals. This requires students to have good insights and to generate creative solutions. Therefore, DT can help

students think like designers, confront difficulties, and solve complex problems in schools, companies, and daily life (Brown and Wyatt 2010). The
five steps of design thinking are defined by David Kelley in IDEO (2001):

(DT 1) empathize: identify the target users and collect information about the
users to solve their problems from their views.

(DT 2) define needs: recognize the needs of your users and define the scope of
the problems.

(DT 3) ideate: think about how to solve problems and brainstorm for creative
solutions.

(DT 4) prototype: use tools to build representations of your idea and make
prototypes.

(DT 5) test: test the idea to the user, discuss whether the problem is solved, and
get feedback.

The designers should go through repeated testing and iterations so
that the final product can be closer to the needs of users.

SITUATIVE COMPUTATIONAL THINKING

In classroom practices, there are two major types of instructional designs
for CT projects but not limited to their variations. One type is to assign
small project tasks to students to design creative products with the given
resources; the other type is to provide big problem scenarios so the students have to come up with creative solutions. The earlier works more
like project-based or task-based learning, such as creating a Scratch game
or creating a functional robot; while the latter is more like theme-based
learning, situated learning, or even game-based learning wherein a larger
learning scenario is presented with a more complex problem. The first
type is practiced more in schools since the latter takes more designing
effort and time for both teachers and students.

Based on the aforementioned statements of either case, we see a need
for a learning model where CT is applied in conjunction with interdisciplinary learning. We defined this as "situative CT" (Shih 2020). As Wing
(2008) states, CT is conceptualizing, not programming; and CT is a way
that humans think, not computers. Therefore, situative CT is further
described as having the following four features that are described as distinct from specific and generic CT (Shih 2020): a) it is a combination
of specific CT and generic CT that leads students to write programming
and solve problems at the same time; b) it is practiced in the contextual

situation, normally a theme-based scenario that relies on related domain knowledge correspondence; c) CT skills are no longer the learning goals but the tools; and d) problems are situated in the scenario in which students have to respond to problems in context instead of conducting uniform tasks.

Overall, situative CT learning scenarios emphasize the individualistic production in real-life scenarios. In the confined and problematic condition, students solve contextual problems by accessing the appropriate resources from limited options and generating learning strategies to be used for solving problems. The class should give students space for individualistic creation rather than uniform production. The advanced-level learning scenario can be fluid and dynamic rather than static; sometimes presented with social dilemmas, students would have to make up strategic and feasible plans spontaneously to confront the complex problems.

To sufficiently address the instructional practice of situative CT, a framework with more flexible guidelines would work better than traditional manual-oriented step-by-step instructions. Design thinking is a process that helps designers systematically observe, analyze, extract, and apply these techniques to solve problems in innovative ways with human concerns (Brown 2009). It uses an interdisciplinary approach (Barak and Assal 2018) by connecting the disciplinary systems to the context of the real world (Breiner, Harkness, Johnson, and Koehler 2012; Honey, Pearson, and Schweingruber 2014). In the leading cases and most worked examples of DT, it is used in producing products that can be mass-produced in the market or widely used by a group of people (e.g., a supermarket shopping cart, a handbag, or a low-cost air-conditioner for classrooms); or it is used to transform organizations or living conditions, working toward solving daily life problems, (e.g., team morale, energy efficiency, or learning effects).

A CLASSROOM IMPLEMENTATION EXAMPLE

INTEGRATION OF DESIGN THINKING AND COMPUTATIONAL THINKING

The integration of any two systems would come in various ways (figure 10.1), such as paralleled (1A), sequenced (1B), intersected (1C), and interwoven (1D) integration models. When two systems are carried out in parallel, they

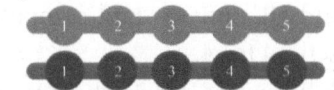

1A. Parallel integration model.

1B. Sequence integration model.

1C. Cross integration model.

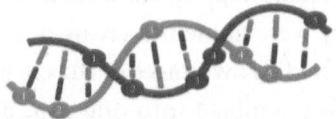

1D. Interweaving integration model.

10.1   Integration models of two systems.

have to perform two works at the same time in a synchronous way (1A). It requires two systems to have similar work patterns and procedures so that they can be coordinated nicely. Nevertheless, it is rare to have two working procedures or conceptual frameworks that can be done together side by side. One of the easiest ways to convey two systems is to do them in consecutive order (1B). However, the sequence integration model is simply performing two things in two different times without interacting with each other. With that said, a cross-integration model is to take turns performing single steps of each system (1C). Attention must be paid to make sure that every step of each system can follow the previous step of the other system so that the complete sequence makes sense. Other than that, the interweaving integration model seems to be the most logical and reasonable way of merging two systems (1D). Some steps from either system would be distinct from each other and have to be performed separately, while some steps are similar so they cross over each other and can be performed together.

COURSE DESIGN AND PRACTICES

With the possible instructional models in mind, an instructional implementation example was conducted in a first grade information technology class in a secondary school in Taiwan (Huang 2020). This example is closer to the practice of project-based learning with the integration of STEM than the dynamic game-based learning scenario. A total of twenty-three students aged thirteen participated this course, with eleven boys (47.8 percent) and twelve girls (52.2 percent). About half of the students had experience in coding with Scratch, and all students had computer information education at the elementary school level.

The goal of this instructional practice aimed to teach students to use technology to work on a collaborative robotics project (situative CT). They were guided to create a combinational project wherein each group devoted its own piece of the creation to the class and all pieces of creations were assembled into one complete functional project (generic CT). Students used computer programming (specific CT) to mobilize the automated mechanism with the connections of robots, sensors, and IoT.

The course was conducted once a week for one class hour with forty-five minutes each. Other than pre- and posttest questionnaires, the teacher used learning sheets to guide students' learning, which were analyzed to ascertain the students' design thinking process. Meanwhile, observations were done to monitor students' learning conditions. The complete research results can be viewed in Huang's study report (2020).

Six groups of students with about four students per group had respective tasks for completing different parts of a simulated automated container terminal (situative CT: theme-based scenario): security gate, container ship, shipping route, crane, container, and cargo truck. In the CT class time, students learned basic programming using Scratch and micro:bit to control movements and functions of the IoT sensors such as ultrasound, motors, buzzers, LED lights, and ultrared light sensors (situative CT: CT skills as tools, not goals).

This attempt treated CT as problem-solving skills and DT as a problem-solving process, therefore conceptually aligning DT in the beginning and the end of the course, with CT in the middle to teach students how to deploy the technologies. The teacher first presented the problem tasks to the students, describing the real-life scenario in which a large amount of cargo ships and trucks come and go and large amounts of cargo are to be transported in the harbor (situative CT: contextual situation and problems). Therefore, an automated container terminal is needed to guide the ships and trucks to go on the designated routes, and the cargo is to be transported automatedly by detections of approaches, weights, directions, and movements of related transportation. At this time, the students searched for information and attempted to not only dissect the problems but also to find creative solutions to the problems with mechanical skills.

The whole teaching model is more similar to the interwoven integration model (figure 10.1: 1D). The teaching concept is to teach students the design thinking process and computational thinking skills at the same time. CT and DT start in a parallel model. The teaching started with the design thinking process of empathizing with the users and defining needs (DT 1 and 2) so that students understood how to approach a problem with a goal in mind. Practice examples of DT used in the course were "how to turn an empty space of classroom into a coffee shop and make a plan about the space design, equipment, menu, etc." Then, decomposition, abstraction, algorithm, as well as pattern recognition abilities (CT 1, 2, 3, and 4) were taught so that the students learn how to dissect the problems and turn the small tasks into programming languages.

After that, three Scratch activities were chosen, such as robot movements, music playing, and interactive sensors, so that the students not only learn how to do programming but also approach tasks in the logic of computer language (CT 5). With the skills at hand, the teacher assigned the class project "Automated Container Terminal" and guided the students to approach their respective tasks in groups. The students then started to go through the design of prototype and development stages of design thinking (DT 3 and 4). Though in the teaching process it looks like CT and DT are sequenced, the students are actually learning two things at the same time since they have to learn CT skills while thinking how the skills can be used in the design of their projects. This is the intersecting point of the two models.

What followed was the process of hands-on work, which was similar to that in many other STEM classes. The only difference was that each group was responsible for a part of the final project, so their design was not only to solve their own problem but also to communicate with other groups and find out the collaborative connections between the groups.

Figure 10.2 shows how the students were working on the collaborative robotics project with computer programming and hands-on maker practices. Their final integrated project showed the efforts of each group where the ship, cargo terminals, crane, and cargo truck can work in the proper sequence and flow well.

From this class project, the students learned from trial and error induced by the problems that failed the mechanism (DT 5). For example,

10.2 Students work on the collaborative robotics project.

the first version of the crane used the L-shaped boom to lift the cargo, but the balsa wood was too thin to handle the heavy cargo, so they changed it into plywood and used an MG996R servo motor to swing the cargo from the ship to the truck. With several iterations, the whole class collaboratively made respective adjustments to fit their pieces into the large project. That was the design thinking spirit this course attempted to deliver.

STUDENT EVALUATIONS

Since the instructional practice focused on both CT and DT, the evaluation also covered both parts. Valid pre- and posttests results were twenty-two copies. The design thinking questionnaire and computational thinking questionnaire each has five corresponding aspects with a total of twenty-three and twenty-five questions, respectively. Both questionnaires are on a five-point Likert scale ranging from 5 as strongly agree to 1 as strongly disagree.

The pretest and posttest of design thinking questionnaire results are shown in table 10.1. The prototype aspect ($t=2.951$, $p=.008$) and test aspect ($t=2.731$, $p=.013$) both reached significant differences. It shows that after the students had the learning experience, their protype and test

Table 10.1   T-test results of design thinking questionnaire

		N	Mean	SD	t value	p value
Empathy	pretest	22	19.14	3.427	1.656	.112
	posttest	22	20.50	3.751		
Define needs	pretest	22	13.82	3.581	1.238	.230
	posttest	22	15.00	3.281		
Ideate	pretest	22	17.91	3.558	1.019	.320
	posttest	22	19.00	3.147		
Prototype	pretest	22	18.00	4.059	2.951	.008**
	posttest	22	20.50	2.972		
Test	pretest	22	14.50	1.896	2.731	.013*
	posttest	22	16.41	2.462		

*p <.05, **p <.01, ***p <.001

abilities improved. But the empathy, define needs, and ideate aspects did not have significant differences. From the observation and after-course reflection with the teacher, there were several in-depth findings and suggestions for conducting this course. First of all, students, especially those in Taiwan, had little training on design projects in a systematic and creative way. The teacher had to provide a lot of guidance in the creation process, so the ideate stage was not totally self-proceeded but required a large amount of teacher assistance. Furthermore, since this Automated Container Terminal project was done from the third-person point of view, its users were difficult to define and the students had a hard time relating, empathizing, and defining needs from a targeted perspective. Thus, the first three aspects did not see significant self-perceived improvements in the questionnaire. On the other hand, since the students were mostly focused on the hands-on projects, their prototype and test aspects had significant improvements.

The pretest and posttest of the computational thinking questionnaire results are shown in table 10.2. The decomposition aspect ($t=2.452$, $p=.023$), algorithm aspect ($t=-2.149$, $p=.043$), and pattern recognition aspect ($t=2.917$, $p=.008$) reached significant differences. This shows that

Table 10.2   T-test results of computational thinking questionnaire

		N	Mean	SD	t value	p value
Decomposition	pretest	22	18.55	3.961	2.452	.023*
	posttest	22	20.64	2.735		
Abstraction	pretest	22	18.14	2.077	1.767	.092
	posttest	22	19.68	2.950		
Algorithm	pretest	22	18.91	2.245	2.149	.043*
	posttest	22	20.64	2.854		
Pattern recognition	pretest	22	18.41	1.681	2.917	.008**
	posttest	22	20.68	2.679		
Generalization	pretest	22	19.86	2.513	1.929	.067
	posttest	22	21.36	2.381		

*p<.05, **p<.01, ***p<.001

after the students had the learning experience, their decomposition, algorithm, and pattern recognition abilities improved. But the abstraction and generalization abilities did not reach significant differences. From the observation and after-course reflection with the teacher, there were several in-depth findings and suggestions for conducting this course. In the teaching process, the abstraction skill was less emphasized since the example given of turning the classroom into a coffee shop did not illustrate the abstraction process and neither did the programming activities with Scratch. Therefore, the students had little idea about what abstraction actually is and could not apply it in their practices. Since the course time is limited and rather short in terms of the semester-long project, the course ended with the completion of the project and summative evaluation without having the chance to approach another similar case for the students to transfer onto. Thus, the generalization skill was not practically used or demonstrated, thus showing insignificant improvement.

From this worked example, several teaching issues were revealed. For example, there is little class time for conducting a large-scale project. The time needed for the students to solve problems is relative to the scale of complexity of the problems. With limited class time, the teachers tend to sacrifice deep thinking for efficiency. Meanwhile, teachers should change their teaching habits to provide guidance instead of instructions. The teaching model should be student-centered rather than teacher-centered and process-based rather than product-based. The evaluation methods should also be transformed into criteria-based instead of test-based. Aspects such as learning behaviors, effectiveness, emotions, and sense of achievement shall be considered as important factors of learning. Evaluation should be more practice-oriented with rubrics, use dynamic observations, and emphasize more the various aspects of creation in the formative way than standardized evaluation in the summative manner.

REVISED INTEGRATION MODEL

As an instruction procedure, it would certainly limit the possibilities of model applications. After this teaching practice of integrating CT and DT, we found that the integration models of type A to type D are not enough to support the work of situative CT. The natures of the concepts of the two systems are not the same. The two models of CT and DT do not correspond to each other step by step, nor do they take care of only the beginning or the end of the production process. They work in a flexible sequence, mostly iterative and sometimes overlapping. Also, since one or more CT skills might be used in every stage of the production, the two models should interweave with each other in more reasonable steps but in a more interchangeable manner. Therefore, type E (figure 10.3) is revealed to be the most feasible integration model.

Since teaching is linear and progresses with time, it is necessary to treat the conceptual framework in the procedural way. The aligned integration model has the benefit of the interwoven integration model (figure 10.1: 1D) while eliminating the rigidity of the cross-integration model (figure 10.1: 1C). It is also important to note that the steps are not locked into what is proposed but should be flexible to accommodate various instructional conditions.

10.3 Aligned integration models of two systems.

Computational thinking encompasses five steps, namely decomposition, abstraction, algorithms, pattern recognition, and generalization (Wing 2008). On the other hand, design thinking involves five stages, namely empathize, define needs, ideate, prototype, and test (d.School 2010). While the former describes more about the techniques used in one or many stages, the later provides a procedural paradigm of carrying out an innovative design.

To give students an overall concept of DT and sufficient understanding of technological techniques related to CT, it is important to schedule DT0 and CT0 stages. In the DT0 stage, the teacher introduces the history, basic concept, practice model, and a worked example practice of design thinking; in the CT0 stage, the teacher teaches various related technologies that might be used in the assigned project. Without going through the CT0 stage, students might generate wild ideas for problem solutions. Their creative design might not be feasible, the difficulty level would be out of control, and resources would be insufficient. Teachers need to choose relevant technological techniques that are extendable and with variations so that students can design their creative products based on the skill foundations taught.

Therefore, guiding students through the design thinking process with the concepts of computational thinking, the suggested aligned integration model offers a procedural flow of the instruction. The teacher presents the problem scenario. Then the students empathize with the people who are involved in the problem, either as active users or passive receivers (DT 1: empathize), and define their needs to focus on the essence of the targeted problems (DT 2: define needs). At this time, both the teacher and the students often blindly jump right into the trial-and-error hands-on work; however, when errors happened too often, students get frustrated and the teacher has to come back to this step to dissect problems. Therefore, it is important to remind the teachers to carefully design the activity in this stage so that problems can be seen in a more constructive way. While facing complex problems, the students should be guided to decompose the big problems into smaller ones (CT 1: decomposition). Seek problems that are comparable to those in previous experiences and search for similar patterns that can be solved with known solutions (CT 2: pattern recognition). At this time, the teachers should find sufficient previous experiences that can provide important guidance and directions. If there aren't enough familiar experiences, a few examples would help to stimulate innovation. The next step is to generate creative solutions that can integrate possible solutions into one (DT 3: ideate). In this step, the question is often about how detailed the idea should be. Suggested principles are to assess whether the design is already functional, concrete, and workable. If it is too conceptual, vague, and imaginative, it will be hard to realize it at the next step. Once the design concept is drafted, the students need to simplify the complexity and try to use more obvious terms or rules to describe the solutions (CT 3: abstraction). This step works more like setting up concrete goals eliciting the main themes of the project. At that point, students attempt to transform the solution into computer logic or using computer languages (CT 4: algorithms). After creating the prototype (DT 4: prototype), students repeat to test the results and adjust the solutions through iterations (DT 5: test). Once the problem is solved, students can use this experience for problems encountered in the future (CT 5: generalization). In class, there might not be a chance to make evident the learning transfer, and it can take longer to reveal the impact of the learning to the students.

This proposed model emphasizes the design thinking process, encourages students to freely design for individualistic creation, and focuses all student teams within and between group projects. It is also important for the teachers to avoid providing direct answers to the students when they encounter difficulties, but to give clues, hints, and updated information and resources so they can solve the problems by themselves. Guiding them to periodically reflect on the DT process will increase their self-awareness and metacognition of every step.

TEACHER TRAINING PROCESS

In Taiwan, there is also a digital divide between urban and rural areas, mostly caused by geographical differences. In the cities, Internet and hardware resources as well as governmental and administrative supports are much more abundant than in the countryside and mountain areas. This also contributes to the insufficiency of teachers in those areas since most teachers prefer to stay in the cities closer to home. The distant schools thus have fewer resources for professional development. The curriculum, consequently, is weaker than those in the cities.

To provide more in-service training to the teachers who work in distant schools, the Taiwanese government invites and assigns professors from the universities as well as leading school teachers to collaborate with K–12 schools to give professional guidance in terms of designing the curriculum and implementing new technologies. By enhancing school principals' leadership, and creating teachers' community of practices, the city schools and rural schools exchange teaching experiences and resources in many ways. Schools, both rural and urban, establish their own school-based curriculum, form their communities of practices, foster teams for STEM, maker, and CT clubs, and sometimes aim to participate in national and even international competitions. With activities that are different from the traditional lecture classrooms, students' learning motivation gets higher, which influences the teachers to move forward with the students. The co-learning cycle becomes more positive.

The whole process is to give teachers a paradigm shift knowing how important CT is, how to integrate CT into their classes, and how to guide students for meaningful practice. There are some rules of thumb for teacher development.

The traditional process of teacher training is to gather the teachers from everywhere and give lectures to them expecting them to put the theory into practice right away. However, it has been found that without constructive guidance and examples to follow, the teachers still do not know how to put conceptual frameworks into practice. Therefore, providing opportunities to experience should be the first step. Just as in the classical saying "to see, to feel, to change," it is better to let the teachers experience the modeled teaching and the effects of change so they have more incentive to learn new things.

In the second step, the key point is to form teacher communities so they can share resources, exchange ideas, and support each other going forward. The goal is to encourage communication between teachers. Meanwhile, conducting needs assessments and providing individualized support can help the teachers do a better job.

The third stage is to build common ground and understanding of the core curriculum concept with the teachers and guide them to customize the design for their own classrooms.

Last, walk together with the teachers. As they put these ideas into practice, be with them and give advice on what they can do. Avoid expecting perfection or unified goals. All schools, teachers, students, and classes are different. Help them through the individualized process.

The double diamond model of design thinking is an iterative process working between divergent and convergent thinking, taking the participants through the discover, define, develop, and deliver stages. It is very similar to the instructional design process while the teachers produce lesson plans. In the same way, it brings the teachers from knowing to doing, from now to new, with insight and goals.

This can provide a model for the teachers when thinking about creating a nurturing scenario. First, emphasize individual creation as well as group collaboration. Avoid one-size-fits-all creation goals, providing assembling manuals, and giving standard prototypes. Then, if possible, extend to a more complex situation and encourage a more creative and dynamic problem-solving process. The teachers can establish a narrative scenario with embedded issues. They can implement clues to get, conflicts to solve, and goals to achieve in order to provide an interaction, creation, and resolution process for the students.

It is an extended goal that through the teacher training process, teachers' motivation and belief can be enhanced and their concerns can be addressed.

CONCLUSIONS

In the era of advanced technology and rapid social change, students are going to face complex problems in the near future. They should have the ability to adapt to the environment in which unpredictable situations might occur. Students should learn to observe, detect, and formulate their own problems along with more collaborative, creative, and critical thinking practices.

From the teachers' perspective, there is the need to integrate existing learning models and form a new one that can be adopted more flexibly. In this chapter, we have proposed that design thinking be integrated into computational thinking classrooms. The proposed teaching model in this chapter should be iterative since the process might vary from case to case due to the various natures of projects.

ACKNOWLEDGMENTS

This study is supported in part by the Ministry of Science and Technology of Taiwan under MOST 104–2628-S-008–002-MY4. I would like to show my appreciation for my research team of several previous studies.

REFERENCES

Angeli, Charoula, and Michail Giannakos. 2020. "Computational Thinking Education: Issues and Challenges." *Computers in Human Behavior* 105: 106185.

Barak, Moshe, and Muhammad Assal. 2018. "Robotics and STEM Learning: Students' Achievements in Assignments According to the P3 Task Taxonomy—Practice, Problem Solving, and Projects." *International Journal of Technology and Design Education* 28, no. 1: 121–144.

Breiner, Jonathan M., Shelly Sheats Harkness, Carla C. Johnson, and Catherine M. Koehler. 2012. "What Is STEM? A Discussion about Conceptions of STEM in Education and Partnerships." *School Science and Mathematics* 112, no. 1: 3–11.

Brown, Tim. 2009. *Change by Design: How Design Thinking Transforms Organizations and Inspires Innovation.* New York: HarperCollins.

Brown, Tim, and Jocelyn Wyatt. 2010. "Design Thinking for Social Innovation." *Development Outreach* 12, no. 1: 29–43.

Chi, Hongmei, and Harsh Jain. 2011. "Teaching Computing to STEM Students via Visualization Tools." *Procedia Computer Science* 4: 1937–1943.

Donovan, M. Suzanne, and John D. Bransford. 2005. *How Students Learn—Science in the Classroom.* Washington, DC: National Academy Press.

d.School. 2010. "An Introduction to Design Thinking Process Guide." https://dschool -old.stanford.edu/sandbox/groups/designresources/wiki/36873/attachments/74.

Fernaeus, Ylva, Mikael Kindborg, and Robert Scholz. 2006. "Rethinking Children's Programming with Contextual Signs." In *Proceedings of the 2006 Conference on Interaction Design and Children,* edited by Kari-Jouko Räihä and Johanna Höysniemi, 121–128. New York: Association for Computing Machinery. https://doi.org/10.1145 /1139073.1139105.

Frymier, Ann Bainbridge, Gary M. Shulman, and Marian Houser. 1996. "The Development of a Learner Empowerment Measure." *Communication Education* 45: 181–199.

Henderson, Peter B., Thomas J. Cortina, and Jeannette M. Wing. 2007. "Computational Thinking." In *Proceedings of the 38th ACM SIGCSE Technical Symposium on Computer Science Education (SIGCSE '07),* 195–196. New York: ACM.

Honey, Margaret, Greg Pearson, and Heidi Schweingruber. 2014. *STEM Integration in K-12 Education: Status, Prospects, and an Agenda for Research.* National Academy of

Engineering; National Research Council. Washington, DC: The National Academies Press.

Huang, Pau-Ching. 2020. *STEM-Based Computational Thinking Course with Design Thinking Method: A Collaborative Project of Automated Container Terminal (ACT)*. Unpublished master's thesis, Tainan City, Taiwan. https://hdl.handle.net/11296/nhmjyv.

Hubwieser, Peter. 2013. The Darmstadt Model: A First Step towards a Research Framework for Computer Science Education in Schools. In *Proceedings of International Conference on Informatics in Schools: Situation, Evolution, and Perspectives*, 1–14. Berlin, Heidelberg: Springer.

Kucuk, Sevda, and Burak Sisman. 2020. "Students' Attitudes towards Robotics and STEM: Differences Based on Gender and Robotics Experience." *International Journal of Child-Computer Interaction* 23–24: 100167.

Ministry of Education. 2014. *Curriculum Guidelines of 12-Year Basic Education General Guidelines*. Taipei: Ministry of Education.

Ministry of Education. 2016. *Draft Curriculum Guidelines of 12-Year Basic Education Guidelines for Science and Technology Courses*. Taipei: Ministry of Education.

National Academy for Educational Research. 2015. *Draft Curriculum Guidelines of 12-Year Basic Education Guidelines*. Taipei: Ministry of Education.

Psycharis, Sarantos. 2013. "Examining the Effect of the Computational Models on Learning Performance, Scientific Reasoning, Epistemic Beliefs and Argumentation: An Implication for the STE Agenda." *Computers & Education* 68: 253–265.

Resnick, Mitchel. 2013. "Learn to Code, Code to Learn." Accessed December 25, 2016. https://www.edsurge.com/news/2013-05-08-learn-to-code-code-to-learn.

Selby, Cynthia, and John Woollard. 2013. "Computational Thinking: The Developing Definition." Accessed April 1, 2014. http://eprints.soton.ac.uk/356481/.

Shih, Ju-Ling. (2022). "Computational Thinking in the Interdisciplinary Robotic Game: The CHARM of STEAM." In *Computational Thinking Education in K-12: Artificial Intelligence Literacy and Physical Computing*, edited by Siu-Cheung Kong and Harold Abelson, 245–269. Cambridge, MA: MIT Press.

Tsai, Ching-Tian, Yen-Hsin Chen, Ming-Lieh Wu, Mei-Kuei Lu, Sheng-Mo Chen, De-Lung Fang, and Yung-Feng Lin. 2011. "A Study Regarding the System Connected the Key Competencies with Individual Academic Field in the Curriculum of K to 12." *National Academy for Educational Research Report* (NAER-99–12-A-1-05-00-2-11). Chiayi County, Taiwan: National Chung Cheng University.

Wing, Jeannette M. 2006. "Computational Thinking." *Communications of the ACM* 49, no. 3: 33–35.

Wing, Jeannette M. 2008. "Computational Thinking and Thinking about Computing." *Philosophical Transactions of the Royal Society of London A: Mathematical, Physical and Engineering Sciences*, 366, no. 1881: 3717–3725.

11

PLETHORA OF SKILLS: A GAME-BASED PLATFORM FOR INTRODUCING AND PRACTICING COMPUTATIONAL PROBLEM-SOLVING

Michal Armoni, Judith Gal-Ezer, David Harel, Rami Marelly, and Smadar Szekely

INTRODUCTION

For over twenty years, Israel's high school system has been following a rather novel program in computer science (CS). The program, which is an elective like biology, chemistry, and physics, was updated periodically according to the development of the discipline. Some of its units have been thoroughly revised and some replaced by new ones. The rationale of this program is to introduce CS as a scientific subject that deals with solving computational problems. The program and its implementation have been described in various studies (e.g., Gal-Ezer et. al. 1995; Gal-Ezer & Harel 1999; and Gal-Ezer & Zur 2004).

More recently, corresponding programs for elementary school (grades 4–6, ages ten to twelve) and middle school (grades 7–9, ages twelve to fourteen) have also been designed and implemented, although in terms of usage some are still in the pilot stage. Such programs should strive to introduce new pedagogical platforms and tools to help teach the discipline's basis, which is new to most of the teachers and learners.

In this paper, we discuss a new approach for teaching young students the basic concepts of CS and endowing them with the relevant problem-solving skills using the innovative, game-based Plethora platform. We also discuss the implementation of the approach in Israel and report on the results of its usage with fourth graders.

Plethora is very different from other platforms with similar goals. First, it builds on a relatively new programming paradigm, *scenario-based programming*, which was launched with the advent of the language of *Live Sequence Charts* (LSC) (Damm & Harel 2001; Harel and Marelly 2003). In the scenario-based approach, a program consists of a set of multimodal scenarios. The execution mechanism follows all scenarios simultaneously, adhering to them all so that any run of the program is legal with respect to the entire set of scenarios. Most scenarios are mandatory whereby if a certain sequence of one or more events occurs, then another sequence must follow it.

The second difference is that Plethora develops computational problem-solving skills without focusing on programming. Students are not expected to code or even to read conventional programs. One of the most interesting of Plethora's underlying principles is to help understand complex systems by changing a given system's *rules of behavior* (i.e., the mandatory scenarios) and seeing the effect of this by viewing the resulting system in action.

Plethora consists of multiple, increasingly difficult, game levels and a level editor with which players can create their own challenges and share them with the Plethora community. The intuitive visual rules are presented in a way that reflects the learner's "mother-tongue." Finally, the Plethora package provides lesson plans with guidance for teachers as well as unplugged activities and real-life examples.

The Israeli Ministry of Education adopted Plethora as a platform in its countrywide cyber competitions and encourages schools to adopt it as an environment for introducing computational problem-solving in elementary school. We have conducted a study, where fourth graders were introduced to Plethora. It shows that Plethora helped the students learn basic concepts of CS and that this experience had the strongest impact when preceded by a short lesson introducing it.

In section 2, we discuss computational problem-solving and its relationship with computational thinking and algorithmic thinking. Section 3 describes the Plethora platform, and in section 4 we discuss teaching and learning computational problem-solving with Plethora. Section 5 discusses the implementation and usage of Plethora in the Israeli school system and our research with the fourth graders. Chapter 6 discusses some thoughts about the future.

COMPUTATIONAL PROBLEM-SOLVING

Computer scientists and computer science practitioners deal with and study computational problem-solving. They solve algorithmic problems by devising appropriate algorithms that are as efficient as possible and establishing their correctness. For problems that cannot be solved algorithmically at all, they try to prove that fact, and for the solvable ones, they attempt to determine inherent levels of complexity. Identifying connections between problems and classifying them by their solvability and complexity are central to their work.

This work is characterized by a set of ideas, thinking patterns, strategies, and skills (e.g., abstraction, nondeterminism, and decomposition) that over the years has been given different names. The first was "algorithmic thinking" (e.g., Knuth 1985). More recently, "computational thinking," originally coined by Papert (1980) and used by Jeannette Wing in her influential article (Wing 2006), has become especially popular. We prefer to emphasize the context in which this set of concepts and skills is used and hence refer to it as *computational problem-solving*, or CPS for short.

The ingredients of this set are not unique to CS. Most of them are useful in other disciplines too, although they may be expressed in different ways, which was the rationale for the call to teach it broadly (Wing 2006). However, as the discipline of CS has developed and matured, the basics of CPS have been enhanced, deepened, and expanded, becoming more powerful and more effective. For example, skills related to data analysis and organization are valuable in many contexts outside CS and were exploited therein long before CS emerged as a discipline. Still, computer scientists have become experts in developing optimal methods of organizing data together with the appropriate actions and manipulations that are to be performed on them. Therefore, teaching CPS from the perspective of CS can provide learners with effective additions to their toolbox, provided they are taught in a generalizable and transferable manner.

Many initiatives for teaching CPS to a young audience utilize designated programming environments and assume that by programming one is exposed to CPS at a level sufficient to enable its use in other contexts. Working in such environments indeed exploits some concepts that are beyond programming per se (e.g., abstraction). However, these are not always explicitly acknowledged and discussed, limiting the usual terminology to the programming act itself and often only to its low-level and technical facet, that is, coding. With such learning, one can hardly expect the students to recognize higher-level ideas and thinking patterns, which is necessary for acquiring them as part and parcel of their toolbox and allowing their transfer to other contexts.

In addition, although many of these environments are based on nonimperative programming (e.g., Scratch, based on the event-driven paradigm, and Alice, based on the object-oriented one), the corresponding educational programs often emphasize the imperative approach. Although some concepts central to these other paradigms are taught (e.g., message passing and events in Scratch), the greater emphasis is on sequential instructions and simple control structures (sometimes also procedures). This inherently limits exposure to CPS to a restricted collection of ideas and habits thereof.

Moreover, in line with Bruner's view on teaching fundamental ideas (Bruner 1960), by failing to explicitly acknowledge paradigm-related issues, one also misses the opportunity to relate to manifestations of the general ideas that are unique to the paradigm at hand, rendering generalization and transfer to other contexts even harder. The classic example of this is the notion of abstraction, essential to CS, which has different manifestations in each of the problem-solving paradigms. An explicit treatment of abstraction, while recognizing similarities and differences among its different flavors, has the potential to promote the acquisition of general abstraction skills as a valuable and transferable tool.

Plethora offers a fresh approach. It is based on a paradigm that is very different from the imperative and procedural paradigms and lends itself naturally to the ability to introduce to young students many of the fundamental concepts of CPS, while emphasizing the phase of devising solutions rather than that of programming them.

PLETHORA

THE PLETHORA GAME

Plethora is an online game-based educational platform. It is organized by topics representing algorithmic concepts, each of which includes numerous levels of varying difficulty.

A *level* is a small system with a set of possible behaviors and is composed of four parts:

1. an *initial state*—a definition of the set of the shapes present on the game screen when the level starts
2. a *goal*—a definition of the set of shapes required to be on the game screen at the end of the level
3. a set of partially specified *activation rules*—when completed, these are to correctly define the behavior of the system represented by this level
4. a *halting rule*—specifies the condition that causes the level to end

The challenge is to complete the partially specified rules so that, when run, they will cause the system to behave correctly: if started in the initial

11.1   Part of the topics screen.

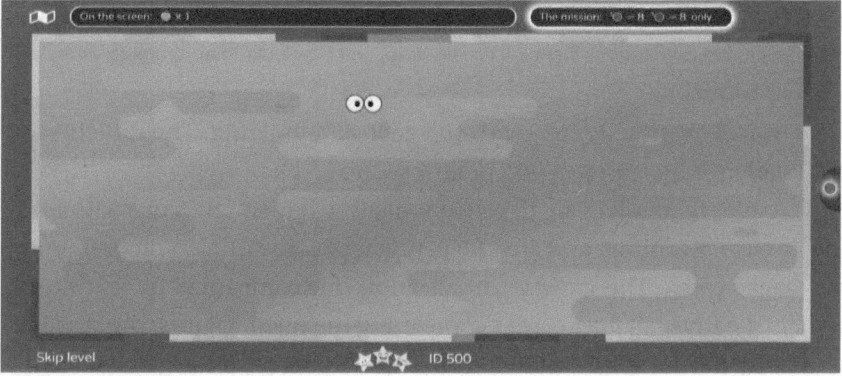

11.2   A simple initial state.

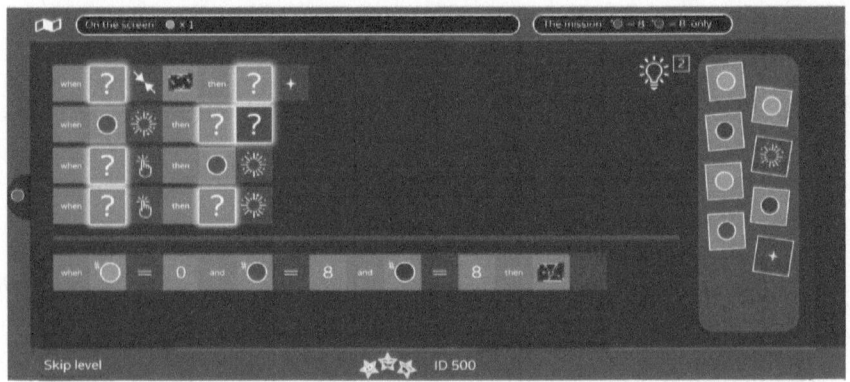

11.3 A level's challenge, calling for filling the empty slots in the rules on the left (denoted by "?") with icons from the "icon tray" on the right, in a way that causes the goal to be achieved.

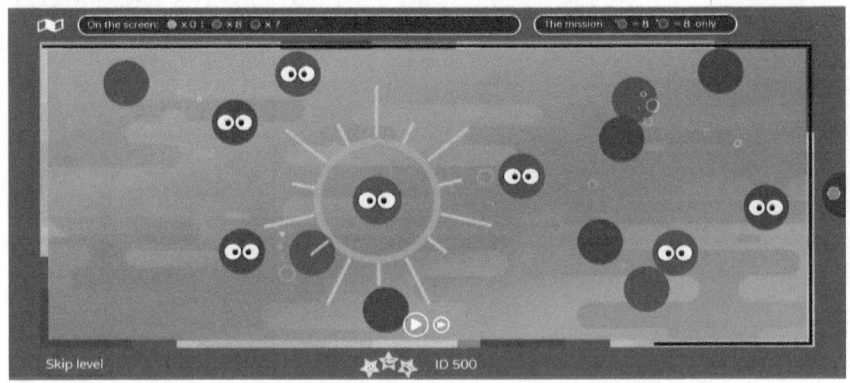

11.4 An execution of the completed set of rules where one can watch the system in action, perform user actions if and when required, and check whether it runs as expected.

state, it will eventually halt with the goal state holding. The difficulty of the challenge increases as one progresses up levels.

The rules themselves are depicted visually, using simple icons with intuitively clear meanings, and are independent of conventional programming language syntax. The graphic visualization is accompanied by text in the students' native language. After completing the missing parts, the students watch the full completed system running/executing on the game screen.

The user can also experiment with completing and activating only some of the rules, observing how these affect others and determining the overall effect on the system's execution.

If the completed set of rules causes the system to reach the goal state, the level ends successfully and the next level unlocks. Otherwise, after a certain fixed running time (currently about forty seconds), the system will halt on its own accord and will ask the student to either continue with the current level's execution or go back to the activation rules and try to solve the challenge in a different way. This iterative process continues until the goal is achieved.

PLETHORA STUDIO

Addressing the challenges Plethora offers, especially the harder ones, requires elaborate computational problem-solving skills and of various cognitive kinds. However, a totally new dimension of creativity is required when asked to create a new challenge from scratch. This calls for assimilating the learned concepts sufficiently well to be able to utilize them in wider contexts, and it is here that Plethora Studio comes in—a level editor with which learners can create their own new challenges and share them with the Plethora community.

Plethora Studio takes students through the structural process of creating a new level, a process that is not unlike actual system design. The initial state and the goal state have to be defined, accompanied by the set of rules that are to lead from the former to the latter.

Once the new level is defined, the user turns it into a challenge by "un-completing" some of the rules, replacing parts of them with the "?" symbol. The challenge can then be shared with schoolmates or with the entire Plethora community.

The combination of a closed approach, where given challenges are to be solved, and an open one, where users create challenges of their own, is of great value. It addresses the two classical ideas of learning through riddles and learning by doing (creating), which are known to be motivating instructional approaches (Schank, Berman, and Macpherson 1999; Delal and Oner 2020).

11.5 Working with Plethora Studio: defining the new level goal state.

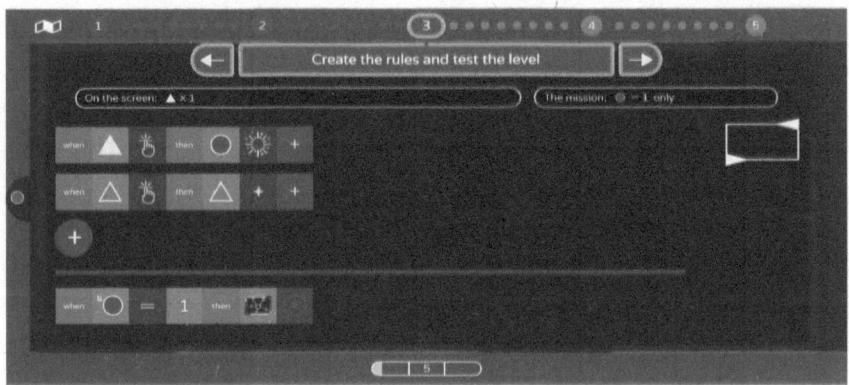

11.6 Working with Plethora Studio: defining the set of rules representing the system's behavior.

PLETHORA AND THE SCENARIO-BASED PARADIGM

Plethora is based on the scenario-based programming paradigm (SBP), originating with the language of *Live Sequence Charts* (LSC) (Damm and Harel 2001), the play-out execution method (Harel and Marelly 2003), and the Play.Go tool (Harel, Maoz, Szekely, and Barkan 2010). It was intended for software and systems engineers, enabling them to intuitively specify and execute complex system behavior without the need for conventional programming. SBP introduces a unique way of specifying requirements, which, rather than being centered on prescribing the system's state-to-state progress, is based on defining scenarios of a systems' behavior, including mandatory, possible, and forbidden ones. It is thus very close to a human's way of thinking about system behavior.

Another important aspect of SBP is that it focuses on the *inter-object* interactions between the system's objects rather than the *intra-object* behavior of each single object. As a result, the focus is on *what* is done in the system rather than on *how* it is accomplished.

Due to the intuitive nature of the SBP and its close alignment with how people express requirements in natural language, we found it to be highly suitable for exposing students to the basics of CPS. Unlike systems for which the shapes are objects that are controlled individually by imperative commands, with Plethora one can describe what should happen from a higher-level perspective without necessarily providing the detailed description of how things are executed.

For example, consider the following rule, which describes a desired abstract behavior.

The rule states that when a yellow star is created, all red circles change their size to be large and all green shapes become blue. However, it does not specify *how* this is achieved (e.g., how do all shapes know that a yellow star was created, or what is the mechanism to notify all red circles that they should change their size?). Hence, a set of rules in Plethora does not look like a set of internal instructions for specific objects but rather a high-level description of system requirements and therefore does not follow the traditional pattern of a program. Indeed, our experience shows that students do not perceive their work with Plethora as programming but as engaging in a challenging game. This makes the CPS class much more attractive.

Plethora covers a variety of subjects, which are also considered important by PISA in the new mathematical framework (PISA 2021). One of the most important of these is *decomposition*, which is actually an expression

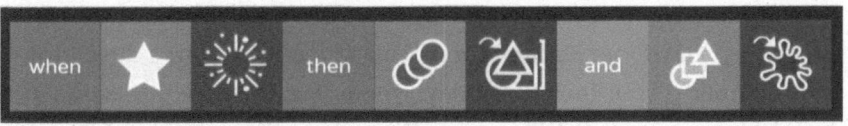

11.7 A Plethora rule.

of algorithmic abstraction, a crucial facet of CPS. Because it follows that scenario-based approach, Plethora's rules represent a natural decomposition of the system's behavior, not its structure. Students are encouraged by the rules to decompose their solution into smaller behavioral fragments that together comprise the overall desired behavior. Hence, we actually have behavioral abstraction.

Plethora supports many concepts inherent to CPS (cause and effect, conditional flow, iterative actions, compound conditions, algorithm halting, multiplicity, and so on), some of which are considered hard to teach in younger grades and are made easier with Plethora. In the next section, we discuss some of these but, due to space limits, not all of them.

Some video clips demonstrating Plethora in action can be found at: https://drive.google.com/open?id=1CEDOIfwxqZe9c5Sh5tLY1dUmrrd2 SIJM.

TEACHING CPS WITH PLETHORA

Our experience shows that playing and practicing alone is not enough for achieving a deep understanding of fundamental concepts. One should also receive carefully constructed and guided explanations and examples thereof. Accordingly, in this section, we elaborate on how we recommend teaching with Plethora.

Plethora's unique approach promotes teaching the foundations of CPS to extend naturally into other aspects of a student's life. This is supported by the use of simple, domain-neutral, abstract geometric shapes but also, significantly, by an accompanying set of lesson plans with which teachers can better explain CPS and show how to use it in day-to-day domain-specific problems. Each of the lesson plans includes an introduction to and explanation of the learned subject, examples from Plethora and from real-life situations, hands-on practice with Plethora, and a summarizing discussion of the student's experience.

The learning process of a computational topic begins with social unplugged classroom activities, exposing students to it in a direct, informal manner. These include games, where, for example, an action performed by one student triggers an action of another (cause and effect) or where several students carry out activities simultaneously (parallelism),

Iterative action lesson structure

Activity	Time
Part 1 – Unplugged activity	10 min
Part 2 – Introduction to iterative action	10 min
Part 3 – Independent activity with Plethora	20 min
Part 4 – Discussion	5 min
	Total: 45 min

11.8   The structure of a lesson taken from Plethora's Iterative Action lesson plan.

and fun quizzes, where students have to guess an object by asking questions about its characterizing properties (objects and properties).

The next step is to introduce the topic in the context of Plethora. The meanings of the icons supporting it are first described according to the semantics established in class in the previous activities, thus further strengthening links between the game and the surrounding reality. The students then start practicing Plethora on their own, or in pairs, solving new challenges as time permits. Some levels are left for the students to continue at home.

During this learning phase, interesting dynamics usually emerge. Some competition between the students obviously exists, but we also witnessed high motivation of students to help those who are experiencing difficulties. Thus, interestingly, competition becomes collaboration, and the teacher becomes a learning facilitator rather than the sole source of knowledge and wisdom.

After practicing on their own, the class ends with a discussion and summary, with students showing how they solved some of the more challenging levels.

This process serves two goals. First, students experience firsthand the fact that the same problem, even limited in size and complexity, may have multiple valid solutions. The discussion can thus address the pros and cons of the solutions proposed (complexity, clarity, and so on). Second, students are required to explain their solutions to the others, so they must be able to articulate the steps taken to build the solution and justify the reasoning behind it. All this is far more demanding than merely solving a Plethora challenge and helps the students learn the principles of CPS, and develop the corresponding skills, on a far deeper level.

After the students are able to solve nontrivial challenges for a certain topic, the learning process takes another step, where they are required to creatively invent their own challenges using Plethora Studio. Here they are exposed to system design issues—planning, adhering to requirements, and so on. They can be asked to create "freestyle" challenges, where they determine the initial state, the final goal, and the difficulty of the level, or to create a challenge that meets certain criteria, for example, being given the initial state and goal state and constraints regarding the number of rules used, the icons that are allowed to appear, and so on. Creating such a challenge, making sure it adheres to the constraints, and then turning it into a problem for their friends is both challenging and fun.

Here are some of the most interesting CPS topics that can be taught with Plethora.

Parallelism: Dealing with systems that exhibit not-necessarily-sequential behavior, where different parts of the behavior can occur simultaneously and thus affect each other in hard-to-predict ways. Parallelism is known to be difficult to teach (Ben-David Kolikant 2004) since it requires the student to consider the system as a whole and take into account the occurrence of simultaneous actions. Nevertheless, being a central facet of real-world complex systems, parallelism is also an important part of CPS.

Being inherent to the scenario-based approach, parallelism is thus a natural part of Plethora, where it shows up in the three forms: (1) the rules are all continuously checked and monitored in parallel; (2) a single trigger can activate multiple rules, causing them to execute in parallel, as shown in figure 11.9; and (3) using the "and" operator, a single rule can specify that two or more events are to occur simultaneously.

11.9 An example of parallel rules.

Plethora distinguishes between "and" (indicating parallel execution) and "and then" (indicating sequentiality). The difference between the two is also emphasized by the execution mechanism of Plethora that uses virtual clock ticks to synchronize the execution of parallel events and delay intervals to enforce sequentiality.

Randomness: The unpredictability in parallelism is enhanced by the randomness inherent to Plethora since the shapes move within the screen in random directions. By experimenting with Plethora, the students become familiar with the effects of randomness and learn to take them into consideration when solving challenges and constructing new ones.

Data abstraction: Plethora rules can refer to abstract shapes by ignoring the values of some of their attributes, thus allowing one to specify the behavior of different objects using a single abstract rule. For example, a rule can say that when any green shape meets any big circle all the red squares change their shape to a star. This way, students understand the concept of multidimensional abstraction without having to explicitly introduce inheritance and polymorphism.

Time: Plethora introduces the concept of real-time and timing constraints using a timer object, which can be reset and queried for its value. Thus, students have to deal with tasks required to be completed within a given time and with actions that can be executed only within an allowed time window.

Recursion: Considered to be one of the most elusive and difficult to teach, recursion occurs when a thing, or a behavior, is defined in terms of itself

11.10 Rules that involve a timer. The first specifies that a click causes a blue square to be created only if no more than four seconds elapsed since the beginning of the execution.

or of its type. The most common kind is procedural recursion, where a procedure can invoke itself. Procedural recursion is based on procedural abstraction, and specifically on the concept of a black box, known to be difficult in its own right (Armoni, Gal-Ezer, and Hazzan 2006). To understand or design a recursive procedure, one has to accept that the procedure does what it is supposed to do before its details are fully known. This difficulty is augmented by the inherent unbounded nature of an entity that includes itself, so to speak. To deal with this, instructors often use the copies model (Kahney 1989), which helps students realize that the invoking and invoked entity are not really the same entity but merely two copies thereof.

Plethora has a simple kind of recursion, also taken from the scenario-based paradigm: a rule can cause the occurrence of the event that triggers it (see figure 11.11). In this way, the baffling self-referential nature of classical recursion is eliminated: there is no explicit (visual or textual) inclusion of the name of a rule inside the rule itself. Furthermore, it is clear that the recurring occurrences of the event are not the same. We thus get the copies model for free. Mutual recursion is also easily represented in Plethora, as depicted in figure 11.12.

Students can experience the power of recursive behavior by executing recursive rules and watching their effect on the game's screen. Figure 11.13, taken from Plethora's lesson plan, illustrates a rule's recursive behavior by

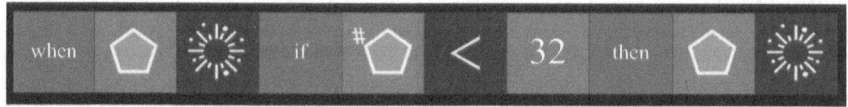

11.11 Recursion: The rule triggers the very same rule by the appearance of one orange pentagram causing the appearance of another.

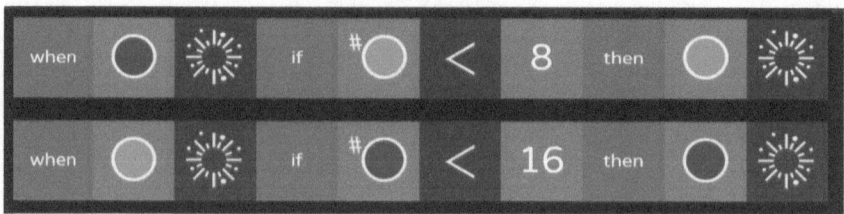

11.12 Mutual recursion.

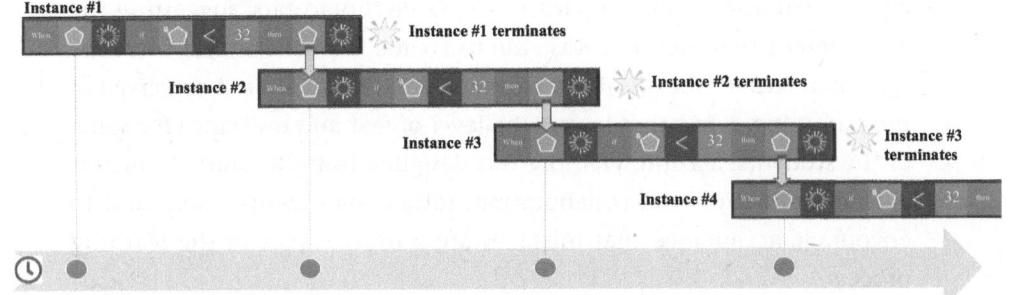

11.13 Multiple instances of a recursive rule, with time flowing from left to right. On the creation of an orange pentagon, a new rule instance is created and activated. This recursive execution will continue until the halting condition is met.

showing how each instance is activated by the effect of a previous instance of the same rule.

PLETHORA IN ISRAEL

ISRAELI CYBER COMPETITION

To encourage students to learn and practice CPS, Israel's Ministry of Education (MoE) organizes an annual event, called the Israeli Cyber Competition (ICC). Students from grades 1 to 12 are exposed for a month to various platforms, according to their age, from which they can choose where to play. The competition is conducted between schools, with each school receiving the points accumulated by its students, normalized by the number of its students.

Plethora was chosen by the MoE as one of the platforms for the 2019 ICC, thus exposing the game to over 55,000 students in grades 5 to 9, countrywide. Following its success, in 2020 the MoE selected Plethora for students in grades 4 to 9, exposing it to an additional 140,000 students.

During the competition, the following topics were introduced via Plethora: cause and effect, user interaction, sequentiality and parallelism, termination conditions, objects and attributes, logical operators, generalization through multiplicity and abstraction, iteration, and recursion. Here are our main insights.

A surprising fact was registered at the junior-high level, where the number of players increased by a factor of six (from three thousand to eighteen thousand) in comparison to pre-Plethora years, suggesting the engagement that Plethora was able to create. We believe that there are two main reasons for this. First, working with Plethora is not perceived as programming, a fact that lowers the level of fear and resistance for some of the students. Second, Plethora was designed from the start to encourage determination and collaboration, rather than competition, and to encourage acceptance that mistakes are a normal part of the learning process.

The fifth and sixth graders played 150 levels in an average time of 10.5 hours, and the seventh to ninth graders played two hundred, more challenging, levels in an average time of 20.5 hours. During the two three-week competitions, students practiced a total of almost twelve million levels, of which eight million levels were completed successfully, and the rest incorrectly.

During the competition's finals, students were exposed to Plethora Studio for the first time. After a short introductory video clip, they managed to create their own levels, which were then handed over to the other competitors to solve. We were very impressed by the enthusiasm of the students in creating their own levels and their excitement in solving levels created by other students.

Finally, one of the most interesting phenomena was gender diversity: 49 percent girls versus 51 percent boys, a balance that CS educators strive to achieve.

11.14 The Israeli Cyber Competition.

IN-CLASS STUDY

We conducted a preliminary in-class study to examine the effects of Plethora on understanding and internalizing the principles of CPS. In particular, we were interested in cause and effect, parallelism, logical conditions, abstraction, and algorithm design. We include here only a very brief description of the research and part of its findings.

The research population included 257 fourth graders from five elementary schools in Israel. They were divided into five groups, each following a different path and then responding to an identical questionnaire. We emphasize that the questionnaire was not composed of Plethora-based questions but rather around a simple problem reminiscent of issues arising in the real world. Specifically, it was about animals in the forest preparing for a party, and the questions referred to various constraints that the animals have to deal with on their way to get to the party on time.

Since the five groups were not of the same size, we chose to examine the pool of *all* answers (the number of questions multiplied by the number of students) and compute the success or failure rate of each group by dividing the number of right or wrong answers of the group by the number of the entire pool of answers, respectively.

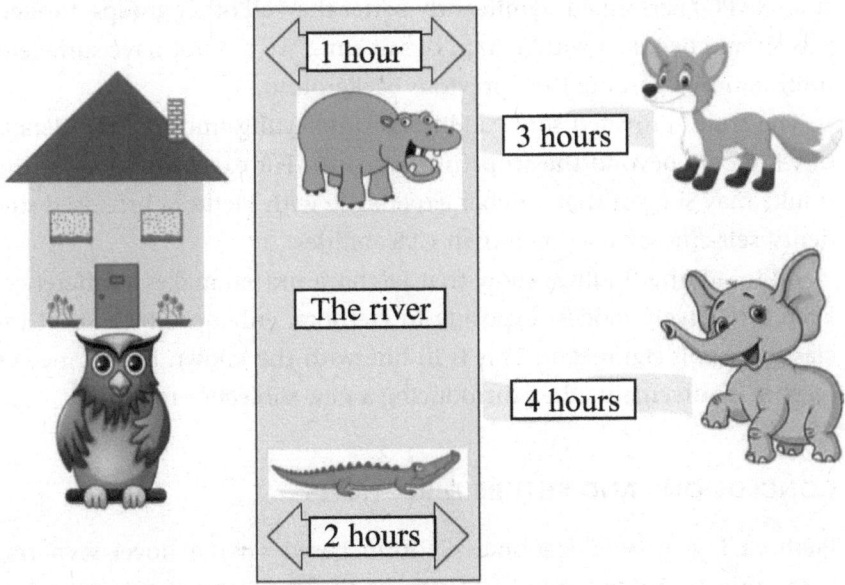

11.15 The problem presented in the questionnaire.

Table 11.1 The five groups participating in the research

Group	No. of Students	Path
Q	51	No prior exposure to Plethora Questionnaire
PQ	82	Experienced Plethora with no guidance Questionnaire
LPQ	68	Attended a Plethora Lesson practiced Plethora Questionnaire
PSQ	24	Practiced Plethora attended a Summary lesson Questionnaire
XPQ	32	Previous eXposure to Plethora Questionnaire

Figure 11.16 shows the findings for all groups. The students in Q (no experience with Plethora) exhibited significantly lower results compared to those in LPQ and PSQ, whereas no significant differences were found between LPQ and PSQ or between PQ and Q. The differences between LPQ and Q, and between PSQ and Q are significant. These results are in line with what we expected: the combination of Plethora with in-class learning contributes to the students' CPS abilities, supporting the learning process described in Section 4.

It is noteworthy that students who experienced Plethora prior to the study (XPQ) performed significantly better than all other groups, though this should be taken with a grain of salt since we did not have sufficient information regarding their previous background.

The study provided some additional interesting findings, the details of which are beyond the scope of this paper. For example, some of the results may suggest that a richer experience with Plethora increased students' self-efficacy regarding their CPS abilities.

All in all, the findings show that Plethora indeed makes a difference. Even a relatively modest exposure to Plethora, enhanced by a short in-class lesson, is significant. This is in line with the known importance of teacher involvement when introducing a new subject or tool.

CONCLUSIONS AND FUTURE DIRECTIONS

Plethora is a new educational platform based on the novel scenario-based programming approach. Utilizing the strengths of this approach

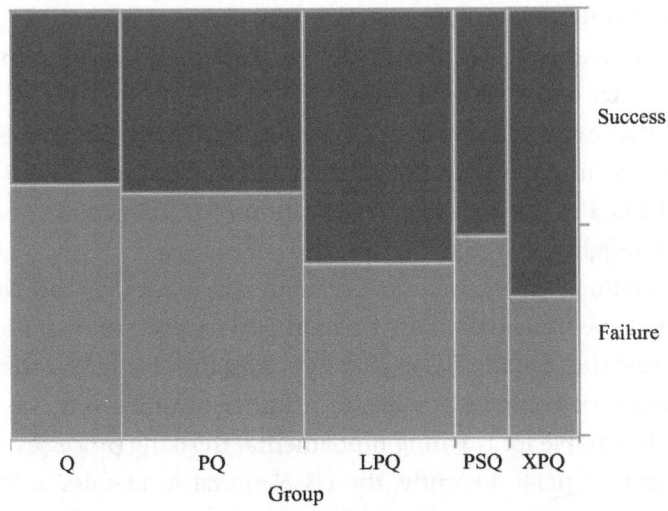

11.16 The success and failure rates for all the groups.

combined with gamification, students design, plan, and understand complex systems and algorithms in an intuitive, fun, and engaging way. The platform, with its associated lesson plans, serves as a powerful tool for developing CPS skills. We envision additional ways to further exploit the potential of Plethora for CPS and other domains.

One of the required skills for the workforce of the future is teamwork, which must be accompanied by the ability to collaborate with peers. Accordingly, our in-class study also examined some aspects of working in pairs. There are, in fact, numerous reports that discuss the advantages and drawbacks of working in pairs (e.g., Tsan et al. 2020; Denner et al. 2014). In the context of Plethora, we found that pairs were able to complete more levels in less time than when working individually. However, we also saw that working in pairs had two drawbacks. One is that it can be done properly only with physical presence, that is, when the pair sit side by side. The second is that a stronger student can be dominant, resulting in the weaker student not really participating (Braught, Wahls, and Eby 2011).

To allow a more constructive way to collaborate, which can be effective also in periods of social distancing and lockdowns, Plethora introduced a teamwork mode, where students connect remotely and solve challenges together. Each student receives only some of the icons and is responsible for completing only some of the rules. Hence, to succeed, they must work together and figure out how each can contribute to the overall process. We plan to investigate the impact of this collaborative mode on the learning experience and its effectiveness.

To further explore the impact of working with Plethora on the understanding and internalization of CPS, we have joined forces with IsraAid organization and the Ministry of Education of Dominica to conduct an extensive longitudinal study in the 2021 school year.

As we claimed earlier, Plethora's mission is to teach the foundations of CPS so that the students can use them in other aspects of their life. Being built around the scenario-based paradigm, and thus being able to describe the behavior of complex systems in a rather intuitive way, we believe Plethora is suitable for teaching fundamental thinking processes relevant to science in general. Recently, the US National Academies of Sciences, Engineering and Medicine (via NRC, the National Research Council) published a framework for K–12 science education. It introduces the notions of *practices*, *crosscutting concepts*, and *core ideas*, emphasizing the need to focus on skills and methods rather than on specific domain content.

This framework is also aligned with the so-called *mechanistic reasoning*, which occurs when students develop explanations that are plausibly aligned with concepts in a particular domain, even if they contain inaccuracies (Southard et al. 2017). This encourages students to draw on their own ideas of explaining phenomena rather than to focus on memorizing the "correct answers" given by their teachers.

There are several similarities between the NRC framework's mechanistic reasoning approach and the process students go through when facing challenges in Plethora. We plan to use the visual infrastructure of Plethora and its scenario-based paradigm and execution mechanism to model various scientific phenomena. The learning process will include challenges to be solved, mainly to get familiar with the scientific terms and basic behaviors, followed by a phase where students build models of the phenomena and try to explain different facts by executing their models.

Introducing the CPS approach for teaching science is a challenge, but we believe that with Plethora and the scenario-based approach, we can achieve the goals of the NRC Framework for K-12 Science Education while at the same time developing students' CPS abilities. Recently, the Israeli

Innovation Authority acknowledged this challenge and has decided to support our research.

ACKNOWLEDGMENTS

We thank Livnat Ben-Hamo, whose master's thesis, supervised by three of the authors, forms the basis of the in-class study subsection, and Shaul Tzionit for the statistics.

REFERENCES

Armoni, Michal, Judith Gal-Ezer, and Orit Hazzan. 2006. "Reductive Thinking in Computer Science." *Computer Science Education* 16, no. 4: 281–301.

Ben-David Kolikant, Yifat. 2004. "Learning Concurrency: Evolution of Students' Understanding of Synchronization." *International Journal of Human Computers Studies* 60, no. 2: 259–284.

Braught, Grant, Tim Wahls, and L. Marlin Eby. 2011. "The Case for Pair Programming in the Computer Science Classroom." *ACM Transactions on Computing Education* 11, no. 1: 1–21.

Bruner, Jerome S. 1960. *The Process of Education*. Boston, MA: Harvard University Press.

Damm, Werner, and David Harel. 2001. "LSCs: Breathing Life into Message Sequence Charts." *Formal Methods in System Design* 19, 1: 45–80. (Preliminary version in *FMOODS'99*.)

Delal, Havva, and Diler Oner. 2020. "Developing Middle School Students' Computational Thinking Skills Using Unplugged Computing Activities." *Informatics in Education* 19, no. 1: 1–13.

Denner, Jill, Linda Werner, Shannon Campe, and Eloy Ortiz. 2014. "Pair Programming: Under What Conditions Is It Advantageous for Middle School Students?" *Journal of Research on Technology in Education* 46, no. 3: 277–296.

Gal-Ezer, Judith, and David Harel. 1999. "Curriculum and Course Syllabi for High-School Computer Science Program." *Computer Science Education* 9, no. 2: 114–147.

Gal-Ezer, Judith, and Ela Zur. 2004. "The Efficiency of Algorithms—Misconceptions." *Computers and Education* 42, no. 3: 215–226.

Gal-Ezer, Judith, Catriel Beeri, David Harel, and Amiram Yehudai. 1995. "A High-School Program in Computer Science." *Computer* 28, no. 10: 73–80.

Harel, David, and Rami Marelly. 2003. *Come, Let's Play: Scenario-Based Programming Using LSCs and the Play-Engine*. Springer-Verlag.

Harel, David, Shahar Maoz, Smadar Szekely, and Daniel Barkan. 2010. "PlayGo: Towards a Comprehensive Tool for Scenario Based Programming." In *Proceedings of the IEEE/ACM 25th Int. Conf. on Automated Software Engineering (ASE)*, Antwerp, Belgium, 359–360.

Kahney, Hank. 1989. "What Do Novice Programmers Know about Recursion?" In *Studying the Novice Programmer*, edited by Elliot Soloway and James C. Spohrer, 315–323. Hillsdale, NJ: L. Erlbaum.

Knuth, Donald E. 1985. "Algorithmic Thinking and Mathematical Thinking." *The American Mathematical Monthly* 92, no 3: 170–181.

Papert, Seymour. 1980. *Mindstorms: Children, Computers, and Powerful Ideas*. New York: Basic Books.

PISA. 2021. "Mathematics Framework." Accessed December 17, 2020. https://www.oecd.org/pisa/sitedocument/PISA-2021-mathematics-framework.pdf.

Schank, Roger, Tamara R. Berman, and Kimberli A. Macpherson. 1999. "Learning by Doing." In *Instructional-Design Theories and Models: A new Paradigm of Instructional Technology*, vol. II, edited by Charle M. Reigeluth, 161–182. Mahwah, NJ: Lawrence Erlbaum Associates.

Southard, Katelyn M., Melissa R. Espindola, Samantha D. Zaepfel, and Molly S. Bolger. 2017. "Generative Mechanistic Explanation Building in Undergraduate Molecular and Cellular Biology." *International Journal of Science Education* 39, no. 13, 1795–1829. https://doi.org/10.1080/09500693.2017.1353713.

Tsan, Jennifer, Jessica Vandenberg, Zarifa Zakaria, Joseph B. Wiggins, Alexander R. Webber, Amanda Bradbury, Collin Lynch, Eric Wiebe, and Kristy Elizabeth Boyer. 2020. "A Comparison of Two Pair Programming Configurations for Upper Elementary Students." In *Proceedings of the 51st ACM Technical Symposium on Computer Science Education*, 346–352.

Wing, Jeannette M. 2006. "Computational Thinking." *Communications of the ACM* 49, no. 3: 33–35.

12

TA TO AI: TINKERING APPROACH TO ARTIFICIAL INTELLIGENCE AND COMPUTATIONAL THINKING EDUCATION IN INDIAN SCHOOLS

Ashutosh Raina, Ronak Jogeshwar, Yudhisther Yadav, and Sridhar Iyer

INTRODUCTION

COMPUTATIONAL THINKING AND ARTIFICIAL INTELLIGENCE

Artificial intelligence (AI) is playing significant roles in our present and will contribute to our future endeavors; hence, there is a need to educate people about it. In his 2020 CTE keynote, Hal Abelson talked about technical skills, social skills, and ethical judgment as essential abilities for understanding and working with AI. Today's computational thinking education (CTE) lacks exposure to skills like statistical methods for data perception and representation; reasoning with creativity and an understanding of models of learning; and natural interactions for developing broad perspectives on social impact, all of which are at the core of AI. A need for developing national guidelines and resource directories for AI education in K–12 has been expressed at many CTE forums.

Computer education in Indian schools has evolved opportunities for developing a lot of projects and applications with computational thinking at its core. Such needs for specialized experience-based education in India have been addressed under the Atal Innovation Mission (AIM) (Atal Innovation Mission, 2020) of NITI Aayog, Govt. of India, through the Atal Tinkering Lab (ATL) initiative. Computational thinking has been developed as an independent module for the ATLs. Further, India's need

for AI education is being addressed at both the state and central government levels. AIM's ATLs in collaboration with the National Association of
Software and Service Companies (NASSCOM 2016) have developed a set
of AI modules. The ATLs are state-of-the-art learning spaces that serve as
a platform for children to explore, play, build, and experience concepts
like AI with these modules, which are in line with the National Education
Policy 2020 (MoE 2020) released by the Ministry of Education, government of India.

The diffusion of AI has enabled a wide range of activities. UN activities
on AI report (Truby 2020) the use of machine learning (ML) techniques for
environmental protection, disaster preparedness, and the development
of cloud-based geospatial solutions for enhanced natural resources management. The proliferation of predictive algorithms and natural language
processing has transformed business processes, including fraud detection,
processing insurance claims, and customer interactions. AI's predictive
capabilities are reducing costs and altering organizational structures. Networked turbines, intelligent power distribution, and automated manufacturing rely on AI technologies (Agrawal, Gans, and Goldfarb 2018).

The structural changes driven by technologies such as AI have both
economic and social implications. The automation of tasks may result
in labor displacement, which may need reskilling/ upskilling for reap-
pointment. The existing AI applications may not have the potential to
replace all tasks conducted by labor in masses but only those tasks that
are routine and noncognitive (Macias-Fernandez and Bisello 2020). It has
been estimated that by 2030, the global market in AI is likely to be in the
range of 15–15.5 trillion dollars out of which India's share will be close
to a trillion (Chen et al. 2016). The next generation of young Indians will
want to challenge the "status quo," and there's no gainsaying; they may
not remain content even with a trillion-dollar share. They will be limited
only by the magnitude of their aspiration.

Apart from increasing the budget allocation for AI, India's government
has plans to improve skills in the area of artificial intelligence, big data,
and robotics. There was an absolute need to revise the education curriculum, not just for technology institutes but also at schools, to include
AI. Moreover, AI needs to be understood as it is socially embedded, interacting with and affecting individuals and communities in myriad ways.

Therefore, AI training should go beyond technology curricula to include social sciences to understand the impacts of AI and contribute to the process of constructing AI algorithms and their assessment. Given its impact, there was a need to introduce AI to students early on to relate to it and associate with its implications to build their foundations over time. To achieve this goal, the following objectives were set:

- Create a fit-for-purpose technology and tinkering-enabled learning and experimentation platform for students to explore various AI dimensions.
- Equip the labs and learning facilitators with necessary tools and learning support to develop student skills, knowledge, and competence in AI. Support will include a holistic curriculum, content, AI DIY Projects (Soft), and so on.

These objectives were addressed by introducing the AI modules in ATLs, a setting that is built to encourage learning and problem-solving by tinkering.

TINKERING AS A VEHICLE TO EXPERIENTIAL LEARNING

Exploration and playful experimentation (play) have been the basis of gaining new knowledge and learning new skills (Resnick and Robinson 2017). In a constructionist approach to learning, building solutions by situating the problem-solving process in an authentic scenario enhances the learning experience (Bransford, Brown, and Cocking 2000). Tinkering is one such practice that includes artefact creation with problem-solving. It has been considered a novice and expert practice that sets it apart from most classroom practices (Danielak, Gupta, and Elby 2014). It does not make tinkering better or worse, but it does make it an authentic professional practice (Berland et al. 2016). Tinkering provides the opportunity to work in a real-time environment with immediate feedback on actions taken, making it a potential means for developing skills by applying and testing one's understanding of concepts. We believe that tinkering with digital platforms and electronic components provides opportunities for exploring concepts and evaluation by applying the AI's concepts to solve problems. Such an explorative hands-on approach supports the development of cognitive and affective skills that support the learning with exploration in higher-education students (Brennan and Resnick 2012). Furthermore, such applied practices have been known to develop motivation and confidence in using newly learned knowledge by overcoming the initial inhibitions and fear of something new.

THE INDIAN TINKERING INITIATIVE FOR K–12: ATAL TINKERING LABS

The Atal Innovation Mission (AIM), housed at NITI Aayog, government of India, sets up state-of-the-art maker-spaces—the ATLs, in schools across the country for adolescent students in sixth to twelfth grade. Children as young as twelve years of age are being introduced to the world of cutting-edge technology with ATL in schools. ATL is the flagship initiative to foster curiosity and encourage students and teachers to experiment, explore, and follow their learning path. It is done by empowering them to think differently about problems and develop innovative solutions by leveraging the latest technology tools including 3D printing, Internet of Things (IoT), robotics, miniaturized electronics, space technology, drone technology, and technology-inspired textiles, and now adding AI to the list. At ATL, students empathize, ideate, design, and prototype using twenty-first-century technologies such as IoT kits, smart electronics, rapid prototyping technologies, 3D printing, and other DIY kits. This has allowed these youngsters to test their ideas and use them to address the grassroots challenges within their local communities. The ATL program's unique pedagogy approach is centered around hands-on learning and making skills (AIM 2020).

Under the ATL scheme, grants-in-aid are provided to schools selected after fulfilling specific requirements of space and workforce and adhering to certain conditions. The aid is for setting up the ATL, which is to be used within a maximum period of five years, including operational and maintenance expenses. As of December 2020, 14,916 schools have been selected to establish ATLs, and approximately 6,500 ATLs have been sanctioned, covering more than 90 percent of India's districts (ATL Schools 2020). These labs have been established in both public and private schools. The majority of them are in coeducational and girls' schools, which are also serving as community hubs of innovation, providing a platform to transform how the youth of the community learns, thinks, ideates, and innovates (NITI Aayog 2018).

To further nurture the school students, the Mentor of Change (MoC) program, a citizen-led national movement, was launched by AIM. Skilled professionals provide pro bono mentoring to young ATL innovators (Atal Innovation Mission 2020). ATL also provides to other sections of the community—including parents, mentors, and other individuals—an opportunity to give life to their ideas. Through frequent community sessions, ATL aims to shape an ecosystem wherein every individual can find solutions to day-to-day personal problems or the society and the country. Such initiatives enhance ATLs' ability and make them ideal for a tinkering-based approach to learning.

ARTIFICIAL INTELLIGENCE MODULES FOR ATL: THE JOURNEY

With ATLs' potential to expose learners to AI, the subject matter expertise was brought in by NASSCOM. It was mutually decided to deliver the AI module in two phases: the AI base module and then the AI step-up module, as the extension of the base module. The initiative's core philosophy was to "trust the learners with their learning path and scaffold them to stand when they fall." The first and most crucial step of the more extensive process was comprehending and understanding the present scenario and expectations of the learners, teachers, and mentors at the school. Many interactions were conducted with school students in the ATLs, their teachers, and mentor communities across the country. Additionally, various models content released or currently deployed across various ATLs were surveyed.

A workgroup from industry and academia was formed to bring in the knowledge, expertise, and experience. The majority of the workgroup constituted tech companies and startups building ed-tech products and services industries using these products. The representatives were practitioners, technology integrators, and educators of AI from organizations like Accenture, Adobe, Amazon, Arm, Bosch, GE, India AI, Microbit, Microsoft, Nvidia, Progilence, SAP, Stempedia, Stemrobo, TechM, Tevatron Tech, Unity, and WIPRO. They contributed by setting expectations, sharing resources, and building activities that were closer to the real-world implementation of AI. The other members of the workgroup were from academia with research experience in technology-enhanced learning, learning environments, and use of educational technology tools and strategies represented by the Interdisciplinary Programme in Ed Tech of the Indian Institute of Technology Bombay. They contributed to the structuring of modules as activity books with an underlying pedagogy based on tinkering. NASSCOM and members from AIM coordinated the activities of the entire workgroup. Group interactions every fortnight gave structure and direction for developing, implementing, and refining the modules. Smaller specialized workgroups contributed to a specific set of outcomes like specific chapters, relevant activities, and examples. The entire content was divided into two modules, the basic and step-up modules. It took around four months each for the base module and the step-up module. NASSCOM hired Progilence to coordinate ground research and to mediate the industry and academia consultation and contributions. They also ensured the alignment of the philosophy, the pedagogical recommendations, learner expectations, and academic curriculum. They were instrumental in instructional design and in publishing, reviewing, and updating the modules.

Based on the ground survey and interactions between the workgroup members, it was realized: (1) the already available content for learning AI is domain- and platform-specific and lacks open exploration, (2) the approach of the available content lacks the opportunities for learning with real-life implementation, and (3) the alignment between the activities of the available learning material with the resources available in the Atal tinkering labs was missing. These aspects were realized to be crucial for an AI learning resource based on tinkering as they help students to apply and evaluate their understanding and expand their exposure to use the concepts in various domains. Moreover, it encourages and engages learners in experiencing AI with activities possible in the spirit of a tinkering lab. Lack of such opportunities could deprive the students from developing an independent capability of working with AI.

To overcome these challenges, it was decided that the modules were to be developed on the basis of progressive formalization, situated in real-life context, argued with analogical scenarios, encourage play by adhering to the norms of tinkering (Resnick and Robinson 2017) and provide scaffolded learning. (These philosophies are discussed in detail in the next section. An overview of their implementation is discussed in the next paragraph, and individual detailed examples have been provided in the next section). As these modules were being developed for a tinker lab, one key aspect driving the design was *tinkerability*, which is ensuring that the learning environment has been designed and set up in such a way that it supports and to some extent encourages the learners to tinker. This can be achieved by incorporating open exploration, fluid experimentation, and immediate feedback (Resnick and Robinson 2017). Fluid experimentation and immediate feedback were achieved by making content interactive with activities, experiments, and simulation platforms guided and encouraged by videos that enable students to work through and learn the various concepts of AI. Open exploration was ensured with activities having a low floor (easy and simple to start without prior knowledge), high ceiling (allows highly complex activities and tasks that require expertise), and wide walls (provides several options to choose from to do the same kind of tasks). A number of platforms were recommended that allowed learners to choose the expertise and challenges they would want to take and guide their learning. The complexity was gradually increased within the chapters of the modules and among the modules. The step-up module built on the concepts and the problems of the base module.

To provide an overview of the flow of a concept in and across the modules, let's look at an example. Let us see how ethics in AI was designed and implemented in the base and step-up module. First, adhering to the progressive formalization, the learning content was broken down into two chapters, unit 7: Ethics on AI in the base module (Link: https://aim.gov.in/Lets_learn_AI_Base_Module.pdf#page177) and unit 2 of the step-up module (Link: https://aim.gov.in/Lets_learn_AI_StepUp_Module.pdf#page=69). The chapter in the first module provides scenarios, asking learners to take a stance with a motive to kindle their thought whereas in the second module the chapter now formally introduces the impact of technology, talking about types of bias and the sources of that bias, with activities and examples as seen in the key learning outcomes shown in figure 12.1.

Now let's look at how the flow introduces the topic of bias in training data in the first module and then extends it to the concept of fairness in

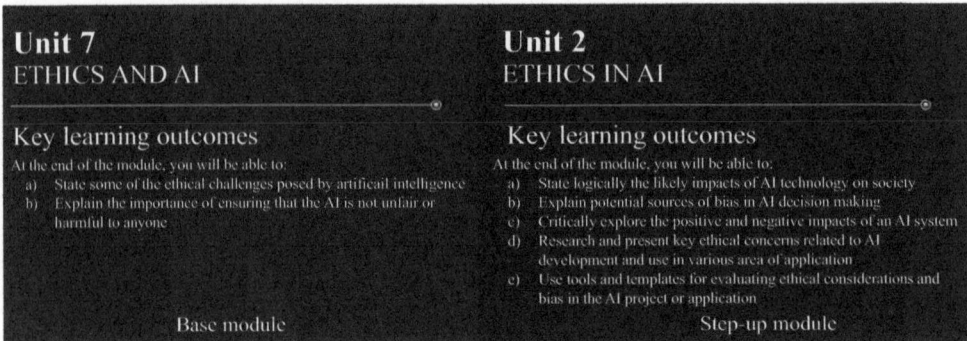

Unit 7
ETHICS AND AI

Key learning outcomes
At the end of the module, you will be able to:
a) State some of the ethical challenges posed by artificial intelligence
b) Explain the importance of ensuring that the AI is not unfair or harmful to anyone

Base module

Unit 2
ETHICS IN AI

Key learning outcomes
At the end of the module, you will be able to:
a) State logically the likely impacts of AI technology on society
b) Explain potential sources of bias in AI decision making
c) Critically explore the positive and negative impacts of an AI system
d) Research and present key ethical concerns related to AI development and use in various area of application
e) Use tools and templates for evaluating ethical considerations and bias in the AI project or application

Step-up module

12.1 Progressive formalization: the variations in the key learning outcomes of the two units on ethics in AI.

the second module. This again is an example of progressive formalization at the topic level. The second scenario in ethical issues of the base module (p. 164) as shown on the left of figure 12.2, puts a student in a real-life situation where they are to realize how their choices of books could have been governed by the corrupt dealing of a shop keeper.

In this case, the learner's empathy helps them relate to a biased recommender system. The step-up module in unit 2 on ethics extends the example of bias to the concept of fairness with more examples, seen in the right side of figure 12.2. Such an approach is also an example of using analogical scenarios situated in a real-life context, which have been extensively used to design the units for ethics in AI. These scenarios where the learners are to take a stand are set as real-life situations to which the learners could relate. To allow learners to experience bias for themselves, the modules further recommend activities like searching for images of doctors and nurses to see the prevalent gender bias of a search engine and then ask them to tinker with search keywords to discover similar bias in various other domains as seen in figure 12.3.

After the activity for gender bias, there is allowance for discussion, and additional activities have been scaffolded with examples in links and videos to enable the learners to explore the topic further. The flow of topics within the module and between the modules is governed by the philosophies to ensure learner engagement with the content.

The base module was specifically devised considering students as young as twelve years of age, with absolutely no prior background in AI.

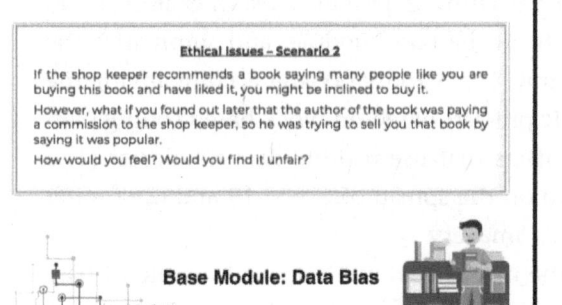

12.2 Analogical scenarios situated in contexts relatable to real life.

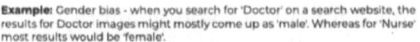

Example: Gender bias - when you search for 'Doctor' on a search website, the results for Doctor images might mostly come up as 'male'. Whereas for 'Nurse' most results would be 'female'.

This shows inherent bias in the search results. Now, if you train your AI models using this dataset, the AI model will be biased too and learn that to be a Doctor, being male is an important criterion.

ACTIVITY

Take one of the above ten ethical concerns and find examples actual or potential and research what impact could this have in different contexts? Some of the contexts could be Healthcare, Banking, Insurance and Finance, Online Matrimony and Dating, Law and Justice, etc. Also, find out what action can be taken by society to address these? Make a presentation to others in your class detailing the above.

12.3 Activities for learners to discover data bias in AI systems.

The objective was to encourage young students' involvement across the country from both rural and urban areas, promoting inclusive learning in all genders. The base module has seven units. Each unit starts with key learning goals and discusses various concepts with examples and various hands-on activities. Each unit also has additional resources for learners to explore based on their interest and a summary of learning from each unit. For easy access, links are provided along with barcodes to be scanned if using a mobile device or tablet. The step-up module provides a medium for learners to expand their knowledge and experiment with their understanding after becoming familiar with the basics of the AI. The module's design allows learners to explore the field on their own at their own pace with these AI modules; hence, a hands-on module was formulated to kindle the learners' thoughts to experiment and experience. The step-up module is made up of eight units. The units follow a similar pattern of key learnings in the beginning, provide a list of examples and activities that have evolved from the base module, and summarize the learning at the end of each unit. The activity booklet accompanying the step-up module has four basic projects based on block-based coding platforms and two advanced projects that use real-life data to address (1) a transportation problem based on the spread of Covid-19 and (2) a water scarcity problem using satellite imagery.

In February of 2020 on the eve of National Science Day, the AI base module was launched across the country. The step-up module was launched in August 2020 on the eve of India's Independence Day. The successor to the base module built on its concepts, practices, and problems. Artificial intelligence is the future, and with constant research and innovations being conducted in the field, there is a need for curious young minds to be aware of it.

LEARNING AI BY TINKERING

UNDERLYING PHILOSOPHY AND PEDAGOGY

The module's board objective is to trust the learners with their learning path and scaffold them to stand when they fall. To ensure the development of the modules with the stated philosophy, *progressive formalization* was chosen to present the concepts gradually, the activities were *situated in a real-life context*, arguments were provided along with *analogical scenarios*, and activities were designed to encourage *play* with the entire learning experience designed around *tinkering* to solve problems which were *scaffolded* with examples, resources, and mentor interventions. All the theories are briefly discussed along with some examples from certain sections of the modules as follows:

Progressive formalization (Bransford, Brown, and Cocking 2000) requires teaching to be designed to encourage students to build on their informal ideas in a gradual, structured manner that enables them to acquire the concepts and procedures of the discipline. The problems, challenges, and examples from the basic concepts have evolved in complexity in the same or a similar domain when the advanced topics are discussed. This approach aims to motivate and get the learners excited about the subject to start with and then retain the excitement by going deeper using the same problem, creating curiosity by increasing complexity, then later addressing the curiosity by making them do activities as the subject matter advances. Let's take the example of the chapter "Learning," which is unit 2 of the base module, and see how progressive formalization works across the chapter. The chapter starts with talking about how we learn with experiences we gain from our senses and then abstracts that process into how a computer would learn based on data it gets from various sources. To situate the concepts, the chapter uses examples and activities from computer vision and natural language processing (NLP). These activities are simple and allow learners to experience computer vision and NLP by (1) allowing the learner to relate the concepts to something substantial and (2) getting the learners motivated about the content and keep them engaged. Now let's look at how progressive formalization of unit 2 of base module governed the design of the units 3, 4, and 5 of the step-up module. Unit 3 of the step-up module about machine learning takes the concepts of base modules that talk about learning with data and introduces the seven-step formal process of machine learning as seen in figure 12.4.

Further the examples of computer vision and NLP from unit 2 of the base module are now unit 4 and unit 5 of the step-up module. These units discuss ML with domain-specific examples and activities that now allow

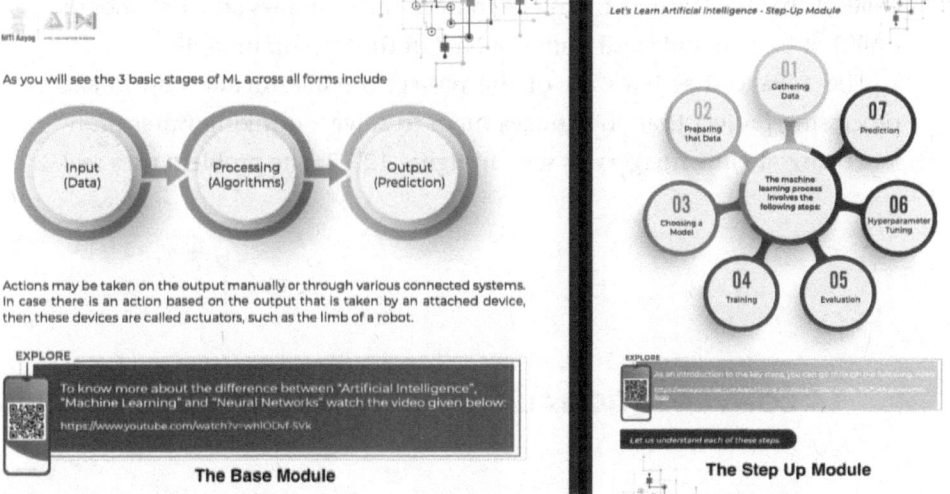

12.4 Formalization of the ML process from the base to step-up module.

learners to build and create the activities they experienced in unit 2 of the base module. In this way, progressive formalization has been designed for concepts along the units they have been discussed in and also in other units across modules.

Learning *situated in a real-life context* (Bransford, Brown, and Cocking 2000) enables a better understanding of abstract concepts by establishing their need in a real-life context using everyday examples. The learners are able to apply and evaluate their understanding with immediate feedback for their actions when based in a real-life context. Activities and projects have been situated in real-life context by providing resources that constitute real-time/real-life data like atmospheric data from open sources and building problems that the learners could relate to. The learners are asked to solve such problems either as thought experiments to get them thinking and/or later by building software- or hardware-based solutions. Activities were designed where the learners are given scenarios and asked about their opinions. Then they challenge their opinions based on conceptual or experiential knowledge. The best example is the projects booklet along with the step-up module. The projects have been divided into basic and advanced. All the projects require the learners to provide real-time data for training ML, and the output is something they can relate to. In the basic set, the conversational interface allows them to build a chatbot that the learners can relate to from the examples and activities of unit 2 of base module and unit 5 of NLP in the step-up module.

The advanced activity 2.1 of the project module includes an image processing project that encourages them to solve a drinking water problem using satellite imagery as seen in figure 12.15. The problem uses real

1.4 CONVERSATIONAL AI INTERFACE

In this project you will build a conversational AI interface to Covid-19 data using 'SAP Conversational AI'

It aims to familiarize you with building conversational AI bots (a.k.a chatbots) with voice and text interfaces to solve a real-world problem.

12.5 Building a conversational interface using AI (chatbot).

satellite images, and the solution to the problem involves finding actual areas covered by waterbodies to compare the coverage with data from a different year. Being able process actual data and do calculations to make an inference about change in surface water coverage impacts the learning of ML and associates it with a substantial outcome.

Analogies have been known to promote learning of the properties of unfamiliar concepts as well as develop new abstractions. Analogy-based scenarios allow the learners to challenge their thought processes to develop argumentation (Gentner and Gentner 1983). Analogical scenarios from multiple perspectives allow learners to associate with challenges that might arise in specific situations, and they have been extensively used to discuss topics like ethics in AI. In both the modules, the chapters on ethics use scenarios as they present some complex situations and then pose questions from multiple perspectives, such as the perspective of the machine, the user, and the programmer. These questions allow the learner to build their ideas and then question them from a different perspective, making them more sensitive about the various challenges associated with AI such as efficiency or ease against privacy and choice. For example, continuing with the example discussed in figure 12.2, the second scenario in the ethical issue of the base module (p. 164): when the learner realizes that his/her choice of books is biased based on a personal interest of the bookkeeper, the learner's empathy from the example scenario helps them relate to a biased recommender system. This example in unit 2 of the step-up module in section 2.2 (p. 53) also discussed in figure 12.3, talks about fairness and shows how biased data could lead to a biased system and have an impact on its users.

Tinkering has been referred to as a playful, experimental, iterative style of engagement in which people are continually reassessing their goals, exploring new paths, and imagining new possibilities (Honey and Kanter 2013). Here *play* has been referred to as experimental play. Play becomes an essential tool for learning in a real-life context as it allows the learners to have a stake in the problem and encourages experimentation with the available resources and one's ideas in the actual problem space with just-in-time feedback that enables reflection (Honey and Kanter 2013). It prepares the learner for real-life scenarios. It also allows one to take multiple perspectives on an action and its impact, which is an essential social skill for developing the mind (Bailey 2002). Tinkering provides a multitude of possible paths taken progressively while situated in a problem space working with immediate feedback. Tinkering as a way of learning focuses on activities built around the concepts that involve using numerous resources to build solutions that allow learners to explore, experiment, and evaluate their understanding of the concepts, in our case in a real-life setting. Tinkering has been known for exploratory learning based on its alignment to several theories of learning. Additionally, the growing availability of design tools has allowed learning through design activities in a constructionist approach (Harel and Papert 1991). Such an approach highlights the importance of young people engaging in learning with the development of external artefacts (Kafai and Resnick 1996) (Roque, Rusk, and Blanton 2013). As discussed previously, learning with tinkering requires the environment and resources to be tinkerable, to support a learners' tinkering ability. Tinkerability was ensured by designing the modules with materials and resources that are known to allow open exploration, fluid experimentation, and immediate feedback. To achieve open exploration, a wide variety of options were made available; for instance, in the activity books, different programming platforms have been provided for the same activity, allowing the learners to choose as per their preference. The same activities could be done on three platforms: ML for kids, Cognimates, and Pictoblox. Various basic activities that start with trying the solutions, like the basic image recognition game in the base module, had been given as an activity of creating similar solutions like building the rock paper scissors game in the project's booklet. Similarly, the basic solutions in the base modules later evolved into complex problems in the step-up module and the activity booklet, which encompass advanced concepts and allow various solution approaches. To support fluid experimentation with quick and easy access to a number of "how to" resources and examples for connecting the activities with concepts have been provided throughout the modules with links and videos available as URLs and barcodes. Immediate feedback has been ensured by using "glass box" platforms and examples that provide the entire solution mechanism, guiding learners to make changes and observe the difference in the changes. Such "glass box" systems allow the learners to observe the impact of the changes they make. Support for tinkering ability was designed into the modules by designing activities for playful experimentation with available AI platforms rather than just giving examples. These activities range from the previously mentioned concept of fairness, discussed with a simple web search activity with the keywords as "doctor" and "nurse" to highlight gender bias, to a more advanced activity of creating and training a system with examples of animals, creating a bias for a specific set. This very example also shows how connections between the concepts of AI and their applications have been designed. Numerous examples that have been listed have evolved as activities with the evolution of the concepts. It was ensured by design that the learners are encouraged to find answers by posing questions and thinking about them. This approach was ensured by providing open-ended problems leaving the learners to figure out the details of the solution scaffolded with questions or examples. The projects book provides the open-ended problem statements in the advanced problem section with links to various resources and an example solution that the students are encouraged to start with and later modify or find a new one based on their understanding.

Scaffolds are like training wheels when one learns to ride a bicycle. As one starts getting the balance right, they are removed (Bransford, Brown, and Cocking 2000). Scaffolds can be technological assistance in the form of digital resources or documents that guide or aid specific processes that may lead to learning, for example, an interactive agent or some design elements of a simulation software. These could be prebuilt or semi-built resources from which the learner could get insights in order to modify them as per their own ideas. Scaffolds could also be in the form of prompts, triggers, or questions from a mentor that aid the learner's thought process. Here, a teacher's role is more of a mentor—one who allows the learners to make decisions on their own, only supporting their thought process and allowing the learners to build their own learning paths. The learners may choose to start by reinventing to gain confidence in the process, which is essential for a learner's motivation and confidence. In today's scenario, the mentor's crucial role is to train the learner to seek information through interaction and investigation rather than provide them with information. The modules have scaffolds in the form of guided activities, quick access links to external resources, and a wide variety of options to choose from. Further, the mentor for change program focuses on training the mentors to guide the learners toward activities for more in-depth exposure to learning. Throughout the modules, the learners go deep into the concepts, but summaries and checks at the end allow them to step back and account for the current topic in the bigger picture of AI.

THE AI MODULES

The entire curriculum has been divided into two modules and a project book for hands-on, real-life challenges. The modules are discussed in detail as follows:

Base module The base module's primary objective is to introduce the learners to the basic concepts of AI and allow them to play with the concepts via activities and experiments aided by the inventory of tech available at the ATLs. The module can be accessed at the link https://aim.gov.in/Lets_learn_AI_Base_Module.pdf or using the QR code shown in the figure. The details of the contents in the base module are as follows:

Chapter 1 on introduction to AI aims at exposing the learner to artificial intelligence and its applications to get them thinking about its utility. The chapter introduces the learners to a range of AI applications on the web using various activities, allowing them to tinker with online models. Figure 12.6 shows an example of such resources that have links and QR codes to access the resources. Simulations are a powerful tool for engagement, especially with new concepts.

These activities and their related concepts are additionally scaffolded using videos and reading materials. The idea is to get learners excited about AI and its applications.

Chapter 2 on learning aims to get the learner to compare and contrast human intelligence and learning within machines. To do so, the chapter uses an online application that introduces machine learning while highlighting its key requirements. Figure 12.7 shows one example of using teachable machines that allow learners to experience how AI can be trained to recognize images. This empowers the learners to experience the training process by doing it on their own. It also introduces students to various ways in which machines learn with respect to human learning

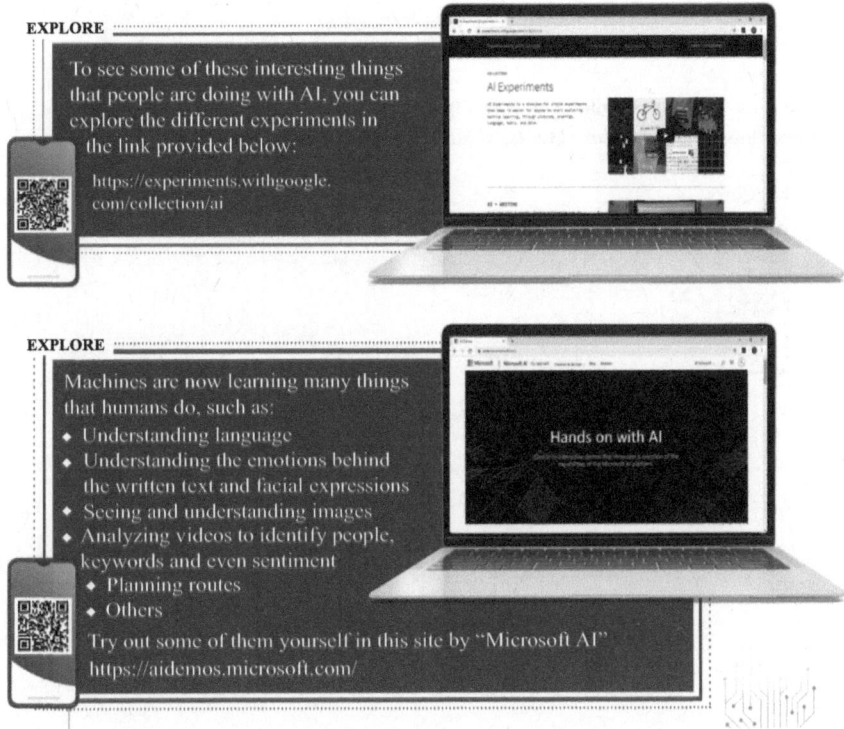

12.6 Actives to explore simulation platforms with quick-access QR codes.

approaches. The examples in this chapter are based on technologies such as speech recognition and computer vision.

Chapter 3 is on data to engage learners with data and data types. Learners do activities to recognize features like centrality. They are provided with different forms of data and try to understand and make simple interpretations from that data. A simple example is shown in figure 12.8 with data in a semiprocessed form from which the learner is to write their interpretations. Learners are exposed to exercises to identify data types as well as ways of capturing data and storing it. The base module has ensured that the learner can differentiate between binary and denary systems and generate binary codes. Scaffolded methods help the learners recognize how binary code systems underpin various data forms like images and sound with applications and examples and exercises.

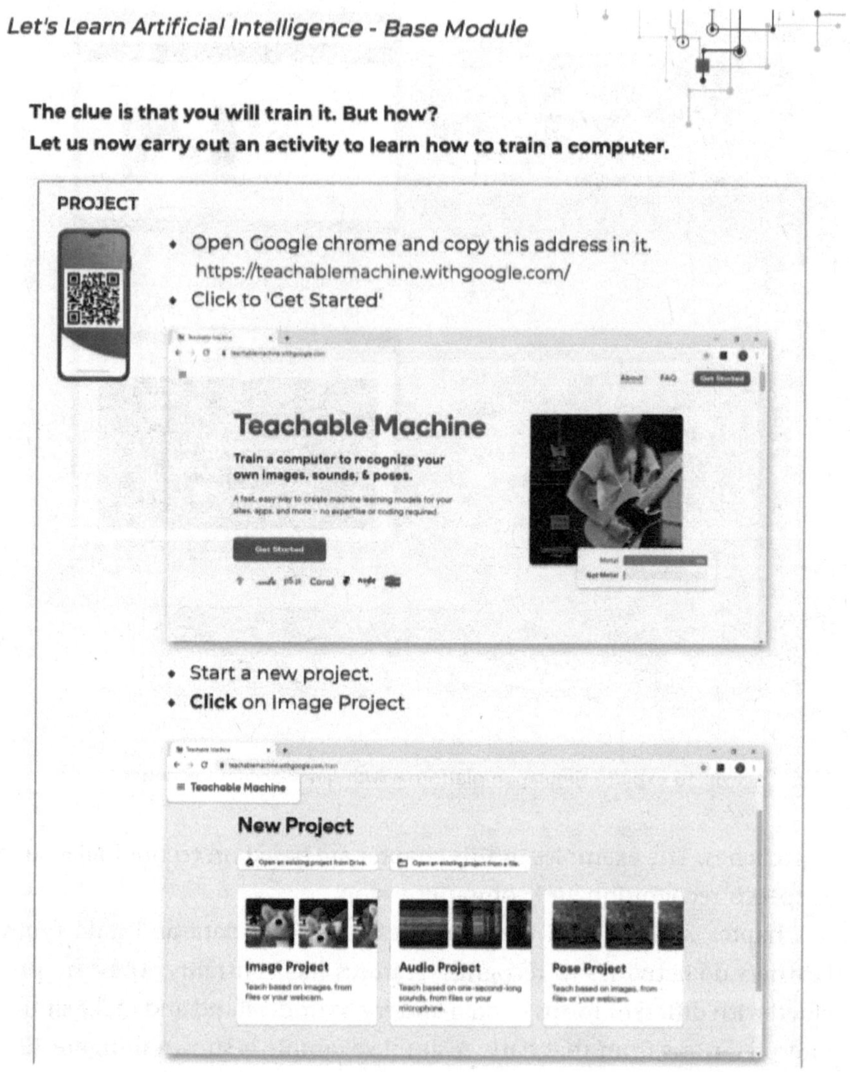

Let's Learn Artificial Intelligence - Base Module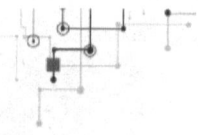

The clue is that you will train it. But how?
Let us now carry out an activity to learn how to train a computer.

PROJECT

- Open Google chrome and copy this address in it.
  https://teachablemachine.withgoogle.com/
- Click to 'Get Started'

Teachable Machine

Train a computer to recognize your
own images, sounds, & poses.

A fast, easy way to create machine learning models for your
sites, apps, and more – no expertise or coding required.

Get Started

♿ ‿ 🐧 Coral ⚡ ♫ 🎵 🎛

- Start a new project.
- **Click** on Image Project

New Project

Open an existing project from Drive.   Open an existing project from a file.

Image Project
Teach based on images, from
files or your webcam.

Audio Project
Teach based on one-second-long
sounds, from files or your
microphone.

Pose Project
Teach based on images, from
files or your webcam.

12.7 Activity that allows the learners to train an AI app to recognize images.

An important takeaway of this chapter highlights the need for data processing capabilities via some simple AI activities.

Chapter 4 on maths and data visualization starts with feature-based classification activity set in a narrative based on likes and dislikes of a pedagogical agent to introduce classification-based AI algorithms. Figure 12.9 shows a stage of a feature selection activity from the narrative of finding the pedagogical agent's favorite shapes. Later, the chapter gets into mathematical

Student Name	Marks in Maths Test
Aarti	YES
Apu	YES
Farah	YES
Anik	MAYBE
Guna	MAYBE
Babu	NO
Devin	NO
Hira	NO
Koel	NO
Total Number of Students: 9	Total YES: 3 Total MAYBE: 2 Total NO: 4

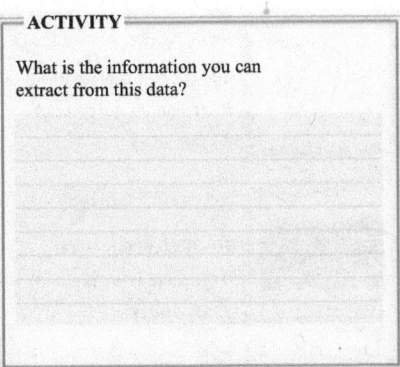

ACTIVITY

What is the information you can extract from this data?

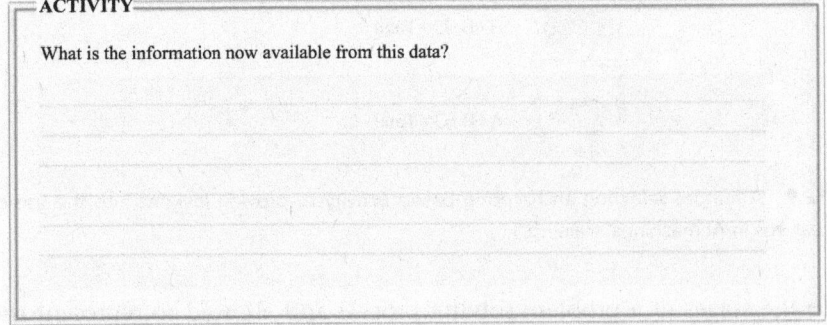

	YELLOW	BLUE	RED	BLACK	GREEN	Total
TRIANGLE	▲	▲	▲	▲		4
CIRCLE	●	●	●			3
SQUARE	■				■	2
Total	3	2	2	1	1	9

ACTIVITY

What is the information now available from this data?

12.8   Activity for data interpretations when data is represented in various forms.

operations and use of introductory algebra, probability, and statistics, linking their application to AI. They also exercise visual data interpretation with basic graphs and play with some online platforms that aid in visualization of data in different types.

Chapter 5 on problem-solving and decision-making starts with a narrative where the learner is trying to solve a problem by recognizing the factors and using an algorithmic process to make decisions that would solve the problem. Through this narrative, the learners are introduced

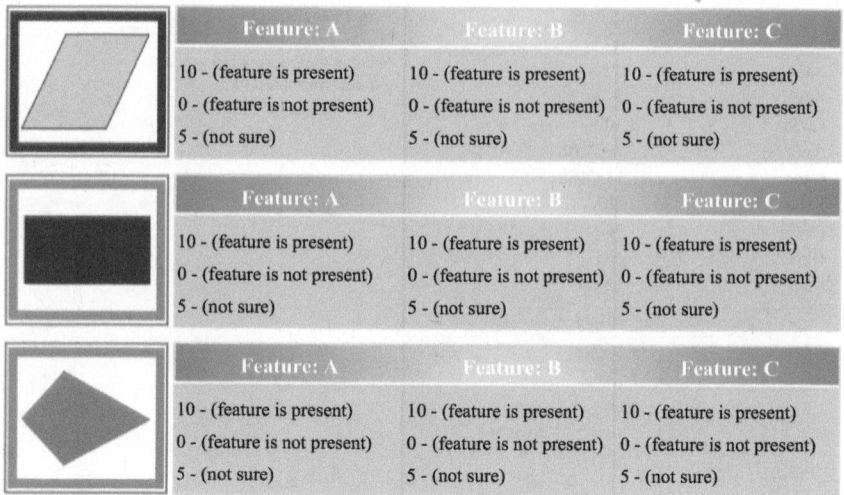

Add up the total marks given to each picture.

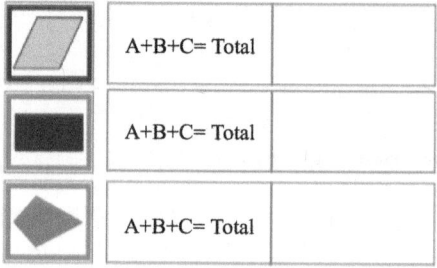

12.9 A feature selection and scoring based activity to provide insights into the use of features in AI machines' training.

to the stages of a problem-solving process and allowed to represent the process in terms of algorithms and pseudocode. The later activities introduce the learner to some basic sort, search algorithms as practice and exposure to algorithmic thinking. Figure 12.10 shows an example where the classification algorithm is taught in terms of classification of a set of animals based on some selected features. The chapter closes with an exercise on the algorithmic representation of a classification algorithm.

Chapter 6 is about introducing learners to programming languages, namely Scratch and Python. The chapter points learners toward prebuilt programs to play with and links to external resources for further

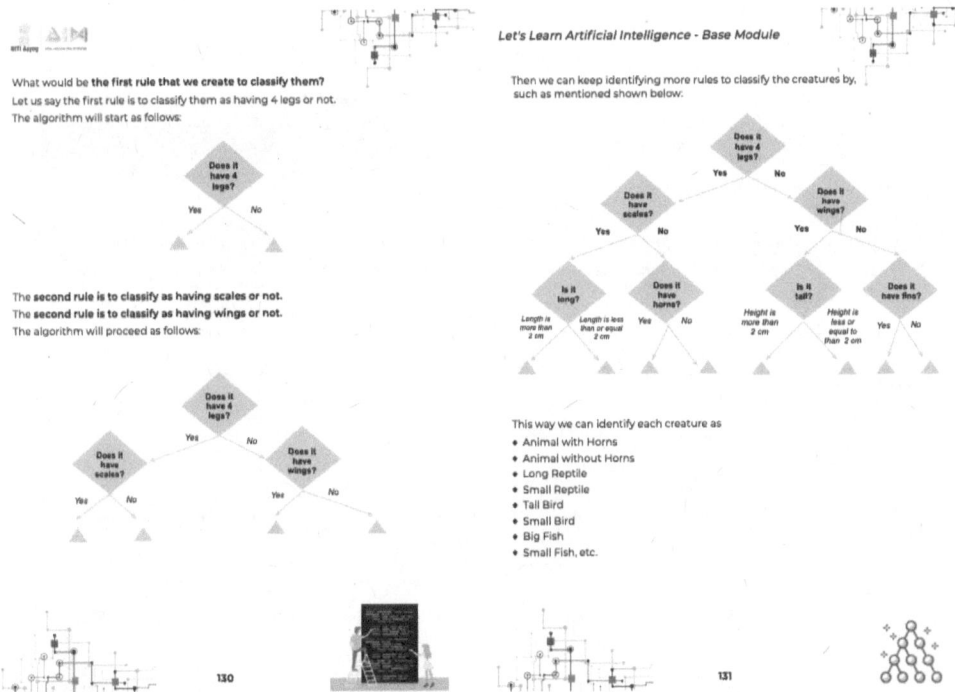

12.10   An example to illustrate classification based on rules represented as a tree.

exploration. The aim is to expose them to two programming basics and allow them to experiment with available codes like a code for smart homes.

Chapter 7 presents the risks and talks about ethics. This is done by posing scenario-based questions to help them reflect how they feel about the issues. The issues cover bias due to data, the concept of privacy, some social dilemmas, and a lot more. The example in figure 12.11 shows individual scenarios given to the learner, enabling them to think about specific challenges that arise with AI. As learners, they would want to be aware of these.

Step-up module   The AI step-up module builds on the base module's activities in terms of concepts and the breadth of applicability. This module is also accompanied by a project booklet. This module is available on the AIM's ATL modules webpage or can be accessed with the link https:// aim.gov.in/Lets_learn_AI_StepUp_Module.pdf. The details of the step-up module are mentioned as follows.

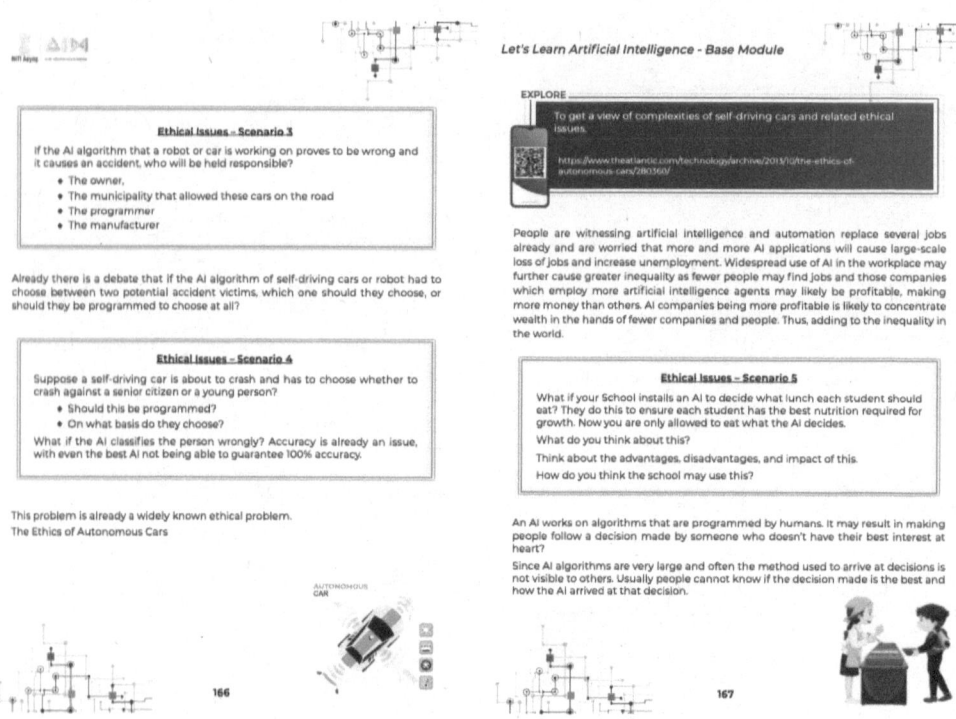

Let's Learn Artificial Intelligence - Base Module

12.11   Examples of scenarios for discussing ethical concerns of AI.

Chapter 1 starts with activities aimed at recalling terms and concepts from the base module. An example of such a recall activity can be seen in figure 12.12. After the activities, there are examples of the use of AI from the domains of health care, astronomy, agriculture, education, entertainment, business, art, data, social media, chatbots, welfare, automobiles, aviation, and deep fakes. The chapter leaves the learners with three ML-based activities like drawing in the air with hands, as shown in figure 12.13, using Scratch-based AI platforms like Pictoblocks (STEMpedia 2020) or ML for kids (IBM ML for kids 2020), where they build programs. The learners are free to choose whichever platform they feel confident or comfortable with.

Chapter 2 builds on the ten considerations of ethics in AI to more structured ethics approach in terms of fairness, robustness, explainability, privacy, and governance. An analogical scenario-based approach has been adopted for delving deep into the topics mentioned previously with

1.1.1 AI BASE MODULE QUIZ

1. Search for seven things that pertain to what AI machines could understand or do from the following word search puzzle. The words may be given horizontally, vertically or diagonally.

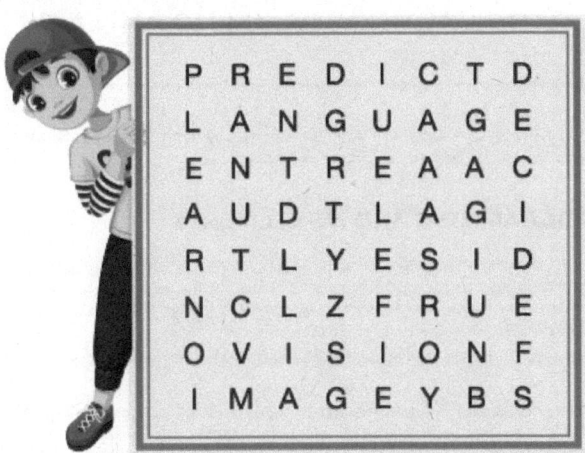

P	R	E	D	I	C	T	D
L	A	N	G	U	A	G	E
E	N	T	R	E	A	A	C
A	U	D	T	L	A	G	I
R	T	L	Y	E	S	I	D
N	C	L	Z	F	R	U	E
O	V	I	S	I	O	N	F
I	M	A	G	E	Y	B	S

12.12 An activity to recall concepts of the base module.

questions to kindle the learner's thought process. Figure 12.14 shows an example of such a scenario that builds the ground for the learners to understand the concept of fairness.

Chapter 3 talks about machine learning, situating it in the realm of artificial intelligence. The chapter walks the learners through the processes of gathering data, preparing the data, choosing an appropriate model, training evaluation, and parameter tuning. Additional information as videos and reading resources are provided with links to example datasets for the learners to experiment with. Further, the chapter talks about neural networks and deep learning with examples. Enough links to external resources are provided for learners to access platforms that offer machine learning as a service.

Chapters 4 and 5 focus in-depth on the domains of NLP and computer vision (CV) used for solving problems. The chapters introduce the learners to these concepts with examples like spam filtering, chatbots, and sentiment analysis for NLP and self-driving cars, facial recognition, and augmented and mixed reality-based applications for CV. The chapters

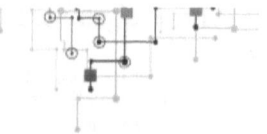

PROJECT: GRAPHICAL PROGRAMMING AND AIR DRAW

In the base module, you were also introduced to graphical programming using Scratch. Let us revisit some of its concepts before we begin our first project.

GRAPHICAL PROGRAMMING AND ITS ELEMENTS

Graphical programming is one of the easiest ways to begin your programming journey. Due to its intuitive interface and programming blocks, it becomes very easy to make games, animations, program robots, and make AI/ML projects. In this module, we will use graphical programming for doing AI-based projects.

Before we jump on to the project, let's first look at the various elements and understand how it works:

Graphical programming (or visual programming) is a method of programming in which the program(s) are written using graphical elements known as blocks. The blocks are like pieces of a jigsaw puzzle that you must put together to solve the puzzle. This approach makes learning to program easy, interactive, and super fun!

You can find more details about the basics of graphical programming at ATL Gaming Module:

https://atlgamingmodule.in/courseInfo/1

Here's a brief overview of any graphical programming platform:

12.13   An ML-based air drawing activity for the learners.

later focus on the working of such examples and finally close with activities that the students can perform on recommended platforms to develop their own text-to-speech engines and chatbots for NLP and facial recognition systems for CV. All the activities discuss the algorithmic approach and encourage the learners to build them on their own on the given platforms without providing step-by-step instructions. However, the activities have been scaffolded by providing reading links and video links for certain technical aspects.

Chapter 6, 7, and 8 focus on building computational skills of managing data and building programs to manage, refine, and prepare data for various AI algorithms with the help of Python libraries. These chapters delve deep into the nitty-gritty of data management with various

3. When it comes to fairness, there is no single approach. What might be fair for one person might not be fair for another. In such cases, the need would be to identify the Fairness Criteria for building the AI system by considering the historical, ethical, social, legal, political and maybe even more scenarios. See picture below to get an understanding of what we mean.

As you can see and understand now, implementing fairness is very important for Artificial Intelligence systems.

And below are some recommended practices that everyone can follow to get a fair AI system in place.

12.14 A scenario to explain the concept of fairness.

forms of databases, using Python's data science libraries, and give the learners basic troubleshooting skills. The last chapter goes into the inner workings of the ML algorithms, allowing the learners to work with supervised, unsupervised, and reinforced learning programming with a few examples from Scratch and Python. These chapters follow a similar activity-based approach and provide pointers toward resources and expertise.

Projects booklet The project booklet was designed as a hands-on manual for learners to try some projects that work on AI. The booklet is available on the AIM's ATL modules webpage or can be accessed with the link https://aim.gov.in/Lets_learn_AI_StepUp_Projects.pdf.

The projects have been divided into two categories. The basic projects use visual programming platforms like adaptations of Scratch for building projects like a rock paper scissors game and a bot using the hardware board called a micro bit available in the ATL labs. The basic platform also includes projects that require the use of AI-based service platforms like Amazon Web Services (AWS) with which the learners build conversational AI interfaces. As for the advanced projects, the book provides two problem statements that are relevant regional problems.

The first one is about identifying water bodies that require conservation as drinking water sources using satellite imagery, as seen in figure 12.15. The second problem is determining the bus routes that can be operated in Bengaluru city while keeping a check on areas that tend to spread Covid-19, as seen in figure 12.16. For both the problems, the data sets are available, and the students have been provided with an algorithmic level of solution to both problems and are encouraged to solve the problems on their own. The learners can also experiment with prebuilt solutions for the given problems, the links to which are also provided. The project book's objective is for the learners to be able to experience the working of AI as part of a solution for a relevant problem, a problem they can associate with.

Let's Learn Artificial Intelligence - Step-UP Module

TOPIC 2 - ADVANCED PROJECTS FOR AI

2.1 IMAGE PROCESSING PROJECT

This project is about identifying most vulnerable fresh water bodies using ML Image Processing techniques.

It aims to make you familiar with training and using image processing models for a real-world problem.

THE PROBLEM

A major problem in India is access to fresh drinking water. Fresh water sources such as rivers and lakes are drying up at an unprecedented pace due to various reasons including climate change, pollution and overuse. Policy makers need data-driven evidence to formulate policies to protect freshwater resources.

Satellite images are available that show the lakes across India. These satellite images are available for a number of years. The task is to identify the increase or decrease in the area of these lakes across time. Lakes that have the highest decrease in area across time can be identified. These lakes can be selected for conservation efforts.

GOALS OF THE PROJECT

At the end of the project, the student will be able to
1. Prepare real-world training and test image datasets
2. Train a deep learning model for image segmentation
3. Apply the model to get predictions
4. Tackle real-world challenges in preparing image datasets, such as, ensuring correct image boundaries and overlapping tiles.

PREREQUISITES

The Student should have
1. Good understanding of image classification and object detection

2. A good understanding of tensorflow

EXPLORE
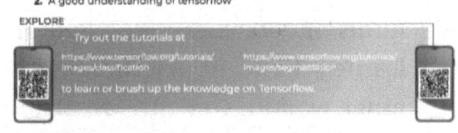

TUTORIALS TO LEARN ABOUT OBJECT DETECTION

The following tutorials and articles can be used by the student to understand how to train and apply object detection models.

https://www.datacamp.com/community/tutorials/object-detection-guide

https://towardsdatascience.com/airplanes-detection-for-satellite-using-faster-rcnn-d307d58353f1

https://medium.com/intel-software-innovators/ship-detection-in-satellite-images-from-scratch-849ccfcc3072

https://www.pyimagesearch.com/2018/05/14/a-gentle-guide-to-deep-learning-object-detection/

https://towardsdatascience.com/data-science-and-satellite-imagery-985229e1cd2f?gi=e93ba19f0a56

https://medium.com/data-from-the-trenches/object-detection-with-deep-learning-on-aerial-imagery-2465078db8a9

DOWNLOADING THE SATELLITE IMAGES

EXPLORE
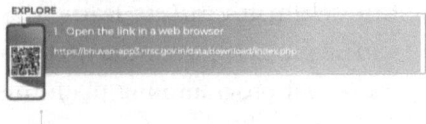

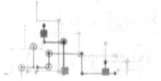

12.15 The drinking water problem statement, prerequisites, and data sources.

PRE-REQUISITES

Data Requirements

Now, that we have the tools setup and running, lets identify the data required for the project. Our objective is to operate buses safely in Bengaluru. So, we need data about all buses operated by Bengaluru's city transportation operator, BMTC. And we also need data about Bengaluru. Note that Bengaluru city is managed by BBMP, a city corporation and the city is divided into WARDs. Each 'ward' has a clearly defined geographic boundary, and has data about its population. So, let's get the following data:

- BMTC data (Routes, Bus Stops, ...)
- BBMP ward details (Zone, Number, Population, Geo Boundaries)
- Ward wise infection data

BMTC data can be found at https://github.com/geohacker/bmtc. Data is available for 2045 routes with information about bus stops, the latitude and longitude of each bus stop, the sequence of stops, etc

BBMP data can be found at https://github.com/openbangalore/bangalore. And, the ward wise population data from https://indikosh.com/city/708740/bruhat-bengaluru-mahanagara-palike

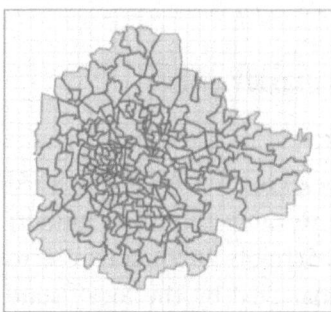

Note, that the BBMP ward and their GEO boundaries are required, so that we can locate all the bus stops within each ward. This is required to create the Origin-Destination flow matrix, for urban mobility which I will explain in a bit

Infection data is not readily available. And the one I could find was at https://indianexpress.com/article/cities/bangalore/covid-19-101-cases-in-bengaluru-so-far-heres-the-list-of-wards-affected-6368503/

The project will use a Python language and libraries.

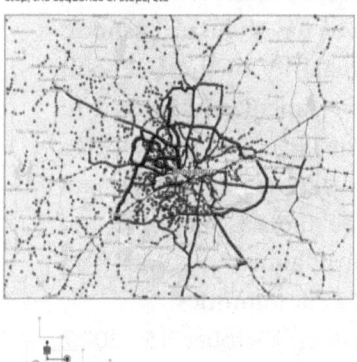

40   41

12.16 The prerequisites and data sources for the Covid-19 public transportation problem.

The learners are free to choose the scaffolding level, which varies from providing them with a general solution and allowing them to build their ideas as they build their version of the solution to starting with the available solutions and then reengineering them as per their understanding and ideas. This ensures a rich experience and exposure to AI-based solutions that are relevant and situated in real life.

These modules are filled with Blackbox and glass box demos, links, and activities from programmable AI platforms, hardware-based activities, a vast repository of videos along with the formal curricula and content, and external resources. These modules aim at enabling learners to experience AI by providing them with the resources to gain skills. The entire initiative's objective is to get the learners excited about AI, scaffold them toward small achievable goals to keep them motivated, and direct them toward extensively available resources and expertise. We believe that motivated and encouraged young learners scaffolded to deal with the initial complexity via these modules in AI can build their learning paths based on their interests, leading them to develop expertise in the domain.

IMPLEMENTATION

REACH

Through the ATL network, the AI base and step-up modules are now accessible to more than two million students in over ten thousand schools (ATL Schools 2020) spread across India. The majority of these schools are managed by the local, state, or central government and located in underserved and rural regions in the country. The larger aim for the ATLs is to provide equal opportunities to the youth to expose them to human-centric design with AI-based strategies to solve India's economic and social impact issues. These modules are currently aimed at students in grades 8–12 irrespective of their schools via ATLs communities.

An app innovation challenge was launched on October 15, 2020, the birth anniversary of India's former president Dr A. P. J Abdul Kalam, observed as World Students Day. It presented an app development opportunity for students to apply their learnings from computational thinking, physical computing, and AI to solve climate protection problems as well as mental health and well-being and other social problems aligned with the UN's sustainable development goals. One thousand teams consisting of two to three students and ATL school teachers/in-charges/mentors developed mobile-based applications addressing challenges in the key focus areas such as ed-tech, health and wellness, agri-tech, Covid-19 response, sports and fitness, mental health and well-being, and gamification (education and entertainment). A total of twenty-one winning teams were selected (three from each focus area) and were offered necessary tools, technologies, resources, and mentorship support to refine their ideas further and enhance their TRL (Technology Readiness Levels). With this first AI contest organized at a school level considering the quality and quantity of the entries, it is clear that the Indian demographic dividend has the talent and the capability to hone the fundamental skills of AI and implement them to solve real-world problems in tandem with the other skills offered by the ATLs.

TEACHER TRAINING

Teachers in ATLs are to take the role of mentors, who are as important as the modules that have been developed. One of the challenges was building the capacity and the confidence of teachers to deliver and support their learners' quest into the realms of AI. With its beneficiaries' assistance, AIM identified such needs like building confidence in the topic and the pedagogical approach of the modules within the school teachers and educators, in collaboration with its partners is conducting free capacity building workshops every three months. Through such courses by professional trainers, the ATL school teachers and educators gain confidence in the technology. They are able to conduct learning sessions for their students within their schools and nearby communities. This has impacted around five thousand teachers who have enabled another fifty thousand students through their guidance (AIM 2021). The training in AI is in tandem with other skill-based modules like app development modules, allowing the teachers to understand the applicability and potential of AI. Even during the lockdowns of 2020 due to Covid-19, as conducting physical sessions within the Atal Tinkering Labs became a challenge, AIM launched a series of learning modules, guided courses, workshops, contests, sessions, and guest lectures under the ATL "Tinker from Home" campaign. Along with the training, a number of mentor resources are available to the mentors (ATL Curriculum 2019).

The training sessions are also designed on a pedagogical approach that supports tinkering. The teachers wear the learner's hat and explore the concepts by performing the modules' activities. As they build this experience, they try to solve challenges that have been posted as projects in the modules, eventually evolving their solutions. The key idea is to expose the teachers to the concepts and technologies using a pedagogical approach; they are encouraged to follow when interacting with their learners. The approach seems to have built confidence in the teachers in supporting the learners in their explorations with AI, as discussed later in the teacher perceptions.

EVALUATION

As the project is still in its first year of inception, currently we just have anecdotal evidence from surveys conducted among the students and mentors from the pilot ATLs to get to know their perceptions when working with the AI modules as out of class activities or project-based activities. The in-depth evaluations of the various factors of the modules and their impact on the schools across the nation are still in the pipeline. In this section, we share the perceptions of a few students and teachers who were part of the pilot launch.

STUDENT PERCEPTION

The hands-on approach has been appreciated and welcomed by all the students. Students who seemed eager to learn about AI said they enjoyed how concepts were aligned with the activities that helped them relate to the concepts' usage. The students who were not sure about AI said they were attracted by the activities and videos of AI applications, which got them interested in the subject, and then they explored further. In addition to the modules, the learners have acknowledged the roles of mentors/teachers as a support structure who can bridge the concepts and their applications using the available resources at the ATLs. A learner from a school in the state of Gujarat said "it (AI modules in ATLs) helped us learn a lot of different things in artificial intelligence. There were teachers who helped us by guiding the uses of sensors, boards, kits and many more things that we have in our lab. We made several different creative projects from ATL lab. We used project kits to upgrade our knowledge and that knowledge we can apply to different projects of different categories we made in ATL like automated farm, automated home. We used artificial intelligence like using a voice assistant in our home automated project that took the project to the next ultimate level."

The course's low floor high ceiling approach seems to have encouraged students who started with the base module, and some teachers helped with projects, allowing students to experiment and explore further. A public sector school learner said, "The design is excellent, especially for beginners to study and understand the basic and advanced concepts in Artificial Intelligence like Computer Vision, Machine Learning." A lot of students said they felt confident about skills associated with working on projects when working with the AI automated home. For instance, a student from a girl's school in the state of Sikkim said "the step-up module provides various tutorials to work on," and then she later said "working with the AI tutorials have helped in developing critical thinking with collaboration." The students seem confident about their learning and the impact of being able to apply their learnings. Learners from different private and public schools in the state of Gujarat said: "To me, it was a gateway to enter into the world of AI and innovations! Gave me a great experience to learn a new skill!" Another student said, "IT IS AMAZING! It gives beginners an in-hand experience from experts to create something innovative and extraordinary!"

The learners' initial perception of the schools that have been a part of the pilot for AI seems promising in terms of what the learners think about the modules. More perception surveys will eventually be rolled out to the entire nation's schools to get rich insights. In-depth studies will be conducted with learners to understand the impact of the various features of the modules.

TEACHER PERCEPTION

Teachers/mentors play an equally important role. This has also been highlighted in the interactions with the learners. The role of a teacher in the pedagogical design of the modules and ATLs, in general, is more of a mentor than just transmission of information. Their contribution goes beyond concepts to nurture the development of cognitive as well as affective skills. The teachers are motivated and trained with conceptual knowledge and its application, as discussed in the teacher training. Teachers have found the AI modules equally interesting and engaging as the students have. They have appreciated the expertise provided by the industry to the mentors. An ATL in-charge from Gujarat said "the AI setup module established in our lab with collaboration with our mentor at . . . is fabulous. It is so easy to use and gives school children the exposure in this field at such a young age. I, as the in-charge am very happy to learn and teach such technology to my students." Even the mentors have appreciated the concept of a high ceiling and wide walls like providing Scratch- and Python-based examples and activities. As another ATL in-charge from a different school in Gujarat said, "they get to know a language which is helpful like python with machine learning." The modules and the teacher training with the industry have helped the teacher's hold on the subject and its application. They feel confident about delivering the content and supporting the learners in this new journey toward AI. A senior teacher and ATL in-charge from Sikkim said, "The AI modules are something that was long-awaited and much needed to guide the students and more so for the teachers. AI, being a contemporary topic in which teachers like us aren't professionally trained, this module has given us the confidence to handle the subject and deliver the right content and support to the children." she later added, "I trust that it will go a long way in shaping the future generation towards the right direction ahead." Lastly, the hands-on approach has also been encouraged by the mentors. As one of them states, "the modules have been excellent tools for the students to maximize the use of AI, enabling them to create and provide improvised solutions to their community problems. It provides a great resource for hands-on learning of various AI-based projects."

Teacher perception has been promising in terms of the intended impact of the elements and the modules' design. As the community grows, the plan is to have more interactions and understand the challenges that are being faced by the mentors and try to find ways of resolving them by making recommendations for the future teacher training programs and industry interactions. The in-depth studies will incorporate the teachers as stakeholders to gather their insights on the learner's interaction with the modules. Surveys from the mentors will be conducted to incorporate their recommendations for future revisions of the modules.

CONCLUSION

The ATLs initiative has been able to lay the foundation for experiential teaching, learning, and problem-solving using project-based active learning strategies and also building a community of learners and makers. Different modules based on different topics aim to exploit the tinkerability of the environment of ATLs. AI is one among such modules that were built ensuring tinkerability and tinkering ability. This was achieved by using a problem-solving and active learning-based approach that encourages learners toward self-exploration of concepts and strategies. Scenarios close to real-life challenges have been used to provide challenges and a lot of examples and prebuilt projects for the learners to experiment with. The modules provide links to various videos and external resources. The extensive reach of the ATLs in all Indian schools has helped the modules to be accessible to the masses. The modules are available in the public domain from the ATL website for anyone to start learning. As mentioned, a few impact studies are in the pipeline that will provide insights into refining the modules' design and pedagogical approach. Secondly, as the teacher/mentor training is carried out, feedback from them will help improve the modules in terms of content and the pedagogical approach. All the recommendations will be taken into account for the next versions. Meanwhile, the hunt for new tools and platforms that can support teaching and learning of AI will help in incorporation of new tech in the next versions of the modules.

REFERENCES

Agrawal, Ajay, Joshua Gans, and Avi Goldfarb. 2018. *Prediction Machines: The Simple Economics of Artificial Intelligence*. Boston, MA: Harvard Business Press.

AIM. 2020. "ATL Handbook." https://aim.gov.in/The_ATL_Handbook.pdf.

AIM. 2021. "ATL Handbook 2.0." https://aim.gov.in/The_ATL_Handbook.pdf.

Atal Innovation Mission. 2020. "AIM." Accessed December 31, 2020. https://aim.gov.in.

ATL Curriculum. 2019. "Atal New India Challenge." https://aim.gov.in/mentor-training-tutorial.

ATL Schools. 2020. "ATL Schools." https://schoolgis.nic.in.

Bailey, Richard. 2002. "Playing Social Chess: Children's Play and Social Intelligence." *Early Years* 22, no. 22, 163–173. https://doi.org/10.1080/09575140220151495.

Berland, Leema K., Christina V. Schwarz, Christina Krist, Lisa Kenyon, Abraham S. Lo, and Brian J. Reiser. 2016. "Epistemologies in Practice: Making Scientific Practices Meaningful for Students." *Journal of Research in Science Teaching* 53, no. 7: 1082–1112.

Bransford, John D., Ann L. Brown, and Rodney R. Cocking. 2000. *How People Learn* (Vol. 11). Washington, DC: National Academy Press.

Brennan, Karen, and Mitchel Resnick. 2012. "New Frameworks for Studying and Assessing the Development of Computational Thinking." In *Proceedings of the 2012 Annual Meeting of the American Educational Research Association (AERA '2012)*, edited by American Educational Research Association, 1–25. https://web.media.mit.edu/~kbrennan/files/Brennan_Resnick_AERA2012_CT.pdf.

Chen, Nicholas, Lau Christensen, Kevin Gallagher, Rosamond Mate, and Greg Rafert. 2016. "Global Economic Impacts Associated with Artificial Intelligence." https://www.analysisgroup.com/globalassets/uploadedfiles/content/insights/publishing/ag_full_report_economic_impact_of_ai.pdf.

Danielak, Brian A., Ayush Gupta, and Andrew Elby. 2014. "Marginalised Identities of Sense-Makers: Reframing Engineering Student Retention." *Journal of Engineering Education* 103, no. 1: 8–44.

Gentner, Dedre, and Donald R. Gentner. 1983. "Flowing Waters or Teeming Crowds: Mental Models of Electricity." In *Mental Models*, edited by Dedre Gentner and Albert L. Stevens, 99–129. Hillsdale, NJ: Lawrence Erlbaum Associates.

Harel, Idit E., and Seymour E. Papert. 1991. *Constructionism*. Norwood, NJ: Ablex Publishing.

Honey, Margaret, and David E. Kanter. 2013. *Design, Make, Play: Growing the Next Generation of STEM Innovators*. New York: Routledge.

Kafai, Yasmin, and Mitchel Resnick. 1996. *Constructionism in Practice: Designing and Thinking*. Mahwah, NJ: Lawrence Erlbaum Associates.

Macias-Fernandez, Enrique, and Martina Bisello. 2020. "A Taxonomy of Tasks for Assessing the Impact of New Technologies on Work." Working paper, Joint Research Centre (Seville site), No. 2020–04.

MoE. 2020. "NEP." https://www.education.gov.in/sites/upload_files/mhrd/files/NEP_Final_English.pdf.

NASSCOM. 2016. "Nasscom. Future Skills." https://futureskills.nasscom.in.

NITI Aayog. 2018. "Strategy for New India." https://niti.gov.in/sites/default/files/2019-01/Strategy_for_New_India_2.pdf.

Resnick, Mitchel, and Ken Robinson. 2017. *Lifelong Kindergarten: Cultivating Creativity through Projects, Passion, Peers, and Play*. Cambridge, MA: MIT Press.

Roque, Ricarose, Natalie Rusk, and Amos Blanton. 2013. *Youth Roles and Leadership in an Online Creative Community*. Cambridge, MA: MIT Lab.

Truby, Jon. 2020. "Governing Artificial Intelligence to Benefit the UN Sustainable Development Goals." *Sustainable Development* 28, no. 4: 946–959.

13

CASE STUDIES OF COMPUTATIONAL THINKING EDUCATION AND ROBOTICS EDUCATION IN CHINA

Su Wang and Jia Li

INTRODUCTION

Computational thinking is the core of IT literacy for students, and robots are key support equipment for advanced manufacturing as well as an important vehicle for improving human lifestyles. Both computational thinking education and robotics education are now highly valued in China. In 2017, computational thinking, as a core disciplinary element of IT curriculum, was written into the *Information Technology Curriculum Standard for General Senior High Schools* (2017 edition) (Ministry of Education of the People's Republic of China 2017, 6) and has been formally introduced into the primary and secondary school curriculum in China since then. Robotics education in China has an even longer history starting in 2003 when it was required to be included in IT curricula (Ministry of Education of the People's Republic of China 2003a, 9–10) and physics curricula for senior high schools (Ministry of Education of the People's Republic of China 2003b, 28), but it had not received sufficient attention until July 2017 when the State Council issued the "Development Plan for the New Generation of Artificial Intelligence" (The State Council of the People's Republic of China 2017), which clearly states that promotions of AI science should be widely carried out and AI-related courses should be offered in primary and secondary schools. Since then, artificial intelligence and robotics have been taught in senior high school IT courses (Ministry of Education of the People's Republic of China 2017, 26–30). The Chinese government is now aware of the importance of computational thinking and robotics education, but policies and curriculum standards for both are developed in a relatively conservative manner.

Chinese academia also attaches great importance to computational thinking education and robotics education, and studies on both have advanced increasingly in depth and extent. In China, the study of computational thinking education in primary and secondary schools formally began in 2012, and since then the number of academic papers in this area has been increasing year by year. Studies in the area mainly fall into three categories: (1) literature reviews that elaborate on the developments and concepts of computational thinking; (2) research on the impact of computational thinking on the IT curriculum; and (3) research on locating computational thinking elements in the IT curriculum (Wang 2017, 21–25). The studies on robotics education in primary and secondary schools in China started roughly in 2002, and they mainly concentrate on four aspects: (1) robotics teaching, including research on teaching vehicles (different types of robots), teaching modes, teaching strategies, learning environment, and so on; (2) educational robots proper, that is, different "robotics technologies" and user experiences; (3) curriculum development and interdisciplinary teaching, that is, to implement robotics education through school-offered robotics courses, STEM-based interdisciplinary education, science education, robotics competitions, and so on; and (4) values and curriculum objectives of robotics education, mainly including cultivating students' innovative thinking, creativity, engineering thinking, problem-solving skills, and so on (Science Promotion Committee of the Chinese Institute of Electronics 2018, 34–35).

A CASE STUDY OF COMPUTATIONAL THINKING EDUCATION

Teaching of computational thinking in primary and secondary schools in China is achieved through two approaches: (1) Computational thinking is integrated into existing subjects: for in-school education, it is integrated mainly into IT courses, sometimes also into other subjects such as mathematics and Chinese (Liu and Zhou 2018, 42); for out-of-school education, it is mainly integrated into programming teaching, robotics education, and so on. (2) Computational thinking is taught through school electives, which mainly include school-based courses developed by the school itself and courses offered by some enterprises. Of these approaches, the integration of computational thinking into IT courses is the area where the largest number of research papers is generated.

Moreover, three vehicles are employed to cultivate computational thinking: (1) screen-based programs, which are primarily written by the use of such programming tools as Scratch, Alice, Python, and so on to create simulations, achieve screen control, or display results on a computer screen; (2) digital tangibles such as circuits and programmable robots, which are mainly used to serve the purpose of developing students' computational thinking through designing and controlling physical robots' activities; and (3) a generic problem-solving approach that focuses on the logic and design of the algorithm and the sequence of steps executed by a computer (Gadanidis et al. 2017, 78). Of all the studies on these three vehicles, cultivating computational thinking through programming has been most studied in China, and Scratch programming in particular is the subject of about 30 percent of all research in this area (Liu and Zhou 2018, 42–43). Scratch programming is relatively easy and can be applied to create interesting games, stories, animations, and other works by combining a series of visual code blocks; therefore, it is suitable for primary and secondary school students to learn. It can not only satisfy students' desire to create and express but also allow them to experience the shift from everyday thinking to computational thinking (Liu and Zhou 2018, 43), improve their analytical and problem-solving abilities, and help develop their mindset. Therefore, this paper chooses to study a case of training students' computational thinking through Scratch programming in primary and secondary schools since it belongs to the most researched and representative type of case studies on computational thinking education.

Following the project-based learning (PBL) approach, a researcher designed an eight-lesson elective Scratch course to develop junior school students' computational thinking (Huang 2019, 36–37). Its instructional design follows six principles: (1) students should propose a project theme first and, under the guidance of the teacher, work out a semi-open project worth exploring; (2) context and problems created for the project should effectively engage students; (3) available resources must be provided for students to understand and solve problems; (4) students are enabled to experience the entire process of learning computational thinking during the project by answering the driving questions that are designed to focus on the process of computational thinking and its application; (5) students are instructed to use Scratch software to produce project works; and (6) evaluation of the project consists of both process evaluation and summative evaluation, with an emphasis on critical thinking. Table 13.1 is an example of the instructional design of a lesson where students learn to create a "drawing board program."

In this case, students' computational thinking is evaluated from two perspectives: (1) an evaluation indicator system is established for Scratch project works, and the computational thinking evaluation is realized in the form of project works evaluation; and (2) A five-dimension computational thinking scale is created covering creative thinking, algorithmic thinking, collaborative learning, critical thinking, and problem-solving. The results showed that the students' computational thinking had improved in general, with problem-solving being the most significantly improved and algorithmic thinking being the least improved. There is a significant gender difference in students' critical and algorithmic thinking but not so in creative thinking, collaborative learning, and problem-solving.

A CASE STUDY OF ROBOTICS EDUCATION

Robotics education has been implemented in China for more than twenty years. The past two decades have witnessed rapid development of robotics education, as exemplified by the change in the types of robots used, from mainly LEGO robots in the past to today's various robots produced by numerous robotics companies. At present, there are principally two ways to implement robotics education in primary and secondary schools in China: (1) to integrate robotics education into other subjects' curriculum, for example, to integrate it into comprehensive practice courses in primary and junior high schools and information technology courses and general courses in senior high schools; and (2) to offer elective courses specialized in robotics education, which may be developed separately or

Table 13.1 Instructional design of creating a drawing board program (Huang 2019, 40–44)

Category	Scratch events	
Requirements	After learning how to apply Broadcast and Brushes, students are required to create a simple drawing board program and a target shooting program with Scratch. Through this problem-solving project, students experience the importance of programming events, further understand what sequence control is, and learn how to achieve sequence control and how to apply it in different programs, comprehend the function of each code block, and grasp how to test and adjust a program in the program-making process.	
Core knowledge and skills	1. Learn how to use the code blocks of "When the character is clicked, broadcast . . ." and "When receiving . . ." in the Events category. 2. Understand the functions of each code block. 3. Apply sequence control to the drawing board program. 4. Transfer skills sequence control to the target shooting program.	
Computational thinking	1. On analyzing the actual needs of painting and entertaining activities, determine the theme of the project and find out the key to problems-solving. 2. Use mind map approach to draw a problem-solving flowchart. 3. Solve the problem of sequence control in the project by producing works with Scratch tools. 4. Be able to apply such problem-solving skills to similar problems (e.g., in the context of the target shooting program).	
Implementation procedures	Instructional activities	Embodiments of computational thinking
(1) Display project cases to engage students in project planning.	Introduction: The drawing board program writing competition is to collect students' works. Introductory activity: Ask students to use the drawing board tool in the computer to draw a simple animal so as to have them familiar with the drawing board.	Identify real-life needs; determine a project theme; analyze task requirements; and search the key to problem-solving.
(2) Discuss and adjust the task list to ensure its availability.	Driving question 1: What functions does a drawing board program need to have? Driving question 2: Can you create a drawing board program using the knowledge you have learned about Scratch? Then, students work out a task list and discuss it in groups. And the teacher provides advice for the items on the list by making suggestions concerning task requirements identification, realization of program functions, choice of characters, and script writing.	Further analyze task requirements; locate the key to problem-solving.

(continued)

Table 13.1   (continued)

Category	Scratch events	
(3) Draw, discuss, and adjust the flowchart in groups to improve its feasibility and availability.	Draw a flowchart for the project based on the adjusted task list to visualize the problem-solving process. Upload the flowchart to the works-display section of Cloud Share and adjust the flowchart after in-group discussions.	Draw a problem-solving flowchart using the mind map approach. Acquire knowledge and skills needed for problem-solving.
(4) Produce project works according to the flowchart.	Design the overall layout of the drawing board, finish the preliminary preparation of the project work (have ready the drawing board background, Brush, Eraser, Paint Board, etc.), write Brush scripts (to enable drawing), Eraser scripts and Paint Board scripts (enable filling of different colors), and display the works.	Develop learners' logical thinking through writing sub-task scripts; train learners' problem-solving skills through creating a program with Scratch tools for visualized programming.
(5) Discuss and evaluate project works in groups.	Students display and self-critique their works, and they evaluate the works through in-group discussions.	Cultivate students' critical thinking.
(6) Transfer of the mindset developed through making the project works	Transfer the knowledge of "sequence control" to the analysis and creation of the target shooting program.	Transfer of computational thinking
(7) Submission of project works	Students submit their works online and the teacher evaluate their works.	
(8) Sustainability of the project	Students think about and answer the question "Are these functions enough for a drawing board program?" and conduct extended research.	Elevate students' computational thinking by asking them the driving questions.

collaboratively by schools and enterprises (Science Promotion Committee of the Chinese Institute of Electronics 2019, 17).

In terms of course content, robotics courses in China fall into two types: those whose main purpose is to teach students scientific knowledge and those to prepare students for robotics competitions and contests (Zhang 2013, 12–13). Robotics competitions play a pivotal role in the development of robotics education in China. In addition to promoting robotics education, such competitions are also used as an important approach to evaluate and assess robotics courses. Robotics competitions, viewed as the most representative approach of robotics education in China, are the result of the application of the "learning-through-competition" pedagogy, and they help to promote robotics education at the primary stage of education (Science Promotion Committee of the Chinese Institute of Electronics 2018, 36–37). Therefore, this paper chooses to analyze a lesson designed for a robotics competition.

China's nationwide robotics competitions for primary and secondary school students include the China Adolescent Robotics Competition (CARC), the RoboCup China Open, and so on. Take CARC as an example. It has many events, including FLL (FIRST LEGO League). Oriented at FLL, a researcher has designed a robotics competition course, as shown in table 13.2.

ANALYSIS AND CONCLUSION

COMPUTATIONAL THINKING EDUCATION

The characteristics of computational thinking education in primary and secondary schools in China can be summarized as follows:

(1) National policy is absent. Over the past decade, education researchers in China have paid increasing attention to computational thinking education, which is reflected in education policies and curriculum standards. Yet, there is still a lack of more specific and comprehensive policies to promote the development of computational thinking education.

(2) Computational thinking education is not widely implemented. Computational thinking is mainly cultivated through IT courses in primary and secondary schools, but information technology courses are valued differently in different regions of China. Generally, first-tier cities pay more attention to IT education, and accordingly the computational thinking education there is better promoted, while IT education and computational thinking education in the central and western regions still need to be vigorously promoted (Ministry of Education of the People's Republic of China 2017).

(3) There are diversified vehicles for computational thinking education. Programming, robotics, problem-solving, and so on can all be

Table 13.2 Course design for FLL Competition on "Food Safety" (Zhang 2013, 34–46)

Principles for course design	(1) Systematized design: to integrate knowledge completely and systematically into teaching contents. (2) Step-by-step process: to break down FLL tasks into several small tasks that are easy to complete. (3) Feasibility: to take into consideration the characteristics of students, the competence of teachers, teaching equipment and other factors when designing the course. (4) Teaching objectives must be clear. (5) Personalization: to pay attention to the difference between students when organizing activities. (6) Full participation: to ensure students' widest participation into and biggest contribution to the competition.
Target students	Upper primary students with some competition experience or related programming knowledge.
Teaching objectives	(1) Knowledge and skills: be able to make small strategic objects according to the competition rules and learn to write and adjust the program. (2) Process and methods: grasp the skills for quick installation and dismantling of small strategic objects according to the competition rules and be familiar with the competition process. (3) Emotions and values: master communication skills, develop teamwork spirit, cultivate the ability to accept both success and failure, and learn to be a person who has a passion for life and is willing to share with others during the competition.
Tools	LEGO robotics components, such as sensors, motors, and various LEGO building blocks, ROBOLAB programming software, USB cables, NXT controllers, computers, and so on.
Teaching procedures	(1) The teacher proposes the theme of the competition. Students try to understand the theme and analyze the tasks including fetching pizzas, sterilizing two vending machines, controlling robots' walk, and supervising storage temperature. (2) Students acquire the theme. After discussion, students decide to combine the following small tasks together: fetching pizzas, sterilizing vending machines, remote transportation by robots, and food preservation. Reason for such combination: the robots can deal with these small tasks in sequence during their walk around the playing field, and they only need to perform two actions by the use of two small strategic objects together without interfering with each other. (3) Students learn the competition rules. The teacher or a student representative reads the rules and requirements of the competition as well as the terms and conditions to win a score. For example, when the robot is preparing to start, it must stay immobile in the starting position; any part of the robot or any item that it is going to move or use must be kept completely within the boundary of the virtual base, with no exception.

Table 13.2 (continued)

	(4) Set up different groups. The FLL competition involves a lot of work such as building robots, programming, planning, and analysis. When grouping students, the teacher should generally consider the talent of each student, taking into consideration their personalities, abilities, and strength in knowledge to ensure that every student is actively involved. (5) Teach students how to make use of the software and some robot-designing skills to lay a foundation for them to build robots. (6) Ask students to design the robot as a group. Based on the competition rules (e.g., time constraints, size requirements for the parts used), the teacher guides the students to generate several reasonable plans through brainstorms. Students build robots, write programs, and decide on the route. This is then followed by continuous debugging and optimization. In this lesson, the robotic design involves design of the robot body, planning of the route, design and assembly of small strategic objects, design of the program, and presentation of the finished product. (7) Select teams. Works and teams are selected according to the design of the robot and students' capability of both independent work and teamwork. (8) Conduct repeated training to improve students' psychological quality. Improve the robot assembled in this lesson (primarily meaning to improve strategic objects and routes, with the robot body remaining essentially unchanged,) and conduct repeated training. Ask students to optimize strategic objects and programs again in coordination with subsequent tasks and to train and adjust the robot in a holistic manner to make sure that each individual task fits perfectly into the time and route.

vehicles for developing computational thinking. The most widely used one in China is Scratch programming. There are already several versions of Scratch textbooks in China, but the way they are compiled is not conducive to developing computational thinking. Scratch textbooks need to be further developed by focusing more on improving computational thinking.

(4) Although diversified teaching strategies are employed, lack of guiding curriculum standards is still true of computational thinking education. Commonly used teaching strategies include the task-driven method, PBL method, "needs analysis+flowchart" method, and so on. In recent years, computational thinking education has been combined with maker education and STEAM education (Liu and Zhou 2018, 43). At the point of instructional design, computational thinking

is dissected and usually different components are identified and taken into consideration, and then different teaching strategies are selected for different teaching procedures to deal with different components correspondingly. Yet some of the teaching cases do not give appropriate consideration as to how to develop students' computational thinking through instructional design but rather take it for granted that students would surely improve their computational thinking through learning the programming features of Scratch. This shows that frontline teachers in China do not have a very clear idea about how to develop computational thinking through Scratch programming, and, what's worse, there is a lack of curriculum standards as well as specific guidance for implementing computational thinking education.

(5) There is no sound evaluation system for computational thinking. Evaluation tools of computational thinking used by schools mainly include computational thinking test papers, questionnaires, and evaluation of works. However, some literature does not include explanation of the reliability and validity of its test papers and questionnaires, indicating the lack of uniform and rigorous standards for evaluation and testing in China.

To sum up, this paper makes the following suggestions for computational thinking education in primary and secondary schools in China: Firstly, more specific national policies and curriculum standards ought to be formulated to lead and guide the implementation of computational thinking education in practice. Secondly, teaching materials and curriculum should be developed with focuses on the cultivation of computational thinking so as to facilitate the popularization of computational thinking education. Lastly, research on the evaluation of computational thinking needs be enhanced, and a sound evaluation system for computational thinking education must be established.

ROBOTICS EDUCATION

The characteristics of robotics education in primary and secondary schools in China can be summarized as follows:

(1) Robotics education is not widely implemented. Although China has issued many educational policies to promote robotics education and

the academic community also attaches great importance to robotics education research, the popularization of robotics education in primary and secondary schools is constrained given the fact that not all schools are able to meet the basic conditions for implementing robotics education, which requires robotics-related technologies, products, platforms, and operating environments to be in place. Currently, public schools in China do not have a sound curriculum system and evaluation system for robotics education, and many primary and secondary schools offer robotics courses through extracurricular activities and interest classes. In addition, robotics education is implemented geographically unevenly in China. In recent years, most schools in Beijing, Shanghai, Guangdong Province, Shandong Province, Hubei Province, and Heilongjiang Province have already offered robotics courses, while other regions are left behind in terms of popularity of robotics education (Zhang 2013, 12).

(2) Diversified types of robots are applied. LEGO robots were the most common type of robots applied at the very early stage of Chinese robotics education. With the development of the robot industry, there is a rapid increase in robot manufacturers in China, and they are able to produce educational robots with various functions and at a wide price range, which helps to promote robotics education (Science Promotion Committee of the Chinese Institute of Electronics 2019, 17–18).

(3) The robotics curriculum is becoming increasingly enriched. There is currently no unified robotics curriculum in China. Robotics courses fall into two types: those whose main purpose is to teach students scientific knowledge and those to prepare students for robotics competitions and contests. The latter is currently the most representative way of implementing robotics education in China (Zhang 2013, 12–13). Robotics competitions have contributed to the advancement of robotics education in China, but they also have some drawbacks. For example, robotics competitions may mislead students about their learning objectives. They prompt robotics companies to develop robot kits for competition purposes only. By purchasing such robot kits, students can easily win a competition. This kind of competition experience is not conducive to developing students' creativity, problem-solving skills, and so on. In recent years, however, robotics

education has begun to embrace new educational concepts such as STEM education and creator education and has become more curricularized and more conducive to developing students' comprehensive literacy (Qiu 2018, 19–35).

(4) There are not enough professional robotics teachers. Robotics is not a compulsory subject in primary and secondary schools, and China has not established a professional training system for STEM teachers. Most robotics teachers do not have an appropriate level of competence as they have not received a formal robotics education (Science Promotion Committee of the Chinese Institute of Electronics 2018, 38). However, robotics teacher training has appeared in some regions, and normal universities have begun to develop related training courses. For example, Hanshan Normal College in Guangdong Province has launched a series of courses to train robotics teachers (Science Promotion Committee of the Chinese Institute of Electronics 2019, 14–16).

In summary, in the future, robotics education in China can be improved in three ways. First of all, an overall robotics education curriculum system should be constructed at the national level, serving as an example for robotics curricula all over the country. Secondly, an evaluation system of robotics education should be established by which students can be evaluated not only in terms of robotics knowledge, such as the basic concepts, structure, functions, and design of robots, but also in terms of their abilities such as logical thinking, problem analysis, problem-solving, comprehensive practice, innovation, and teamwork as well as their truth-seeking spirit (Science Promotion Committee of the Chinese Institute of Electronics 2019, 19). Finally, a specialized training system should be established for cultivating professional robotics teachers to address the shortage of professional robotics teachers in China.

REFERENCES

Gadanidis, George, Janette M. Hughes, Leslee Minniti, and Bethany J. G. White. 2017. "Computational Thinking, Grade 1 Students and the Binomial Theorem." *Digital Experiences in Mathematics Education* 3: 77–96.

Huang, Jin. 2019. "An Experimental Study on Cultivating Junior School Students' Computational Thinking by Using Project Teaching Method—Take 'Scratch Programming' as an Example." Master's diss., Hunan Normal University.

Liu, Lijun, and Xiongjun Zhou. 2018. "A Review of Chinese Studies on Computational Thinking Cultivation in Primary and Secondary Schools." *China Information Technology Education* 11: 38–44.

Ministry of Education of the People's Republic of China. 2003a. "(Abstract of) The Technology Curriculum Standard for General Senior High Schools (Trial)." *Information Technology Education for Primary and Secondary Schools* 5: 9–19.

Ministry of Education of the People's Republic of China. 2003b. *The Physics Curriculum Standard for General Senior High Schools* (Trial). Beijing: People's Education Press. http://www.moe.gov.cn/srcsite/A26/s8001/200303/W020200401347865029995.pdf.

Ministry of Education of the People's Republic of China. 2017. *Information Technology Curriculum Standard for General Senior High Schools* (2017 ed.). Beijing: People's Education Press.

Qiu, Jin. 2018. "The Design and Development of the STEM-Integrated Robot Maker Education Project." Master's diss., Wenzhou University.

Science Promotion Committee of the Chinese Institute of Electronics. 2018. *2018 Report on Robotics Education in Primary and Secondary Schools*. August. http://www.newsstat.cn/education/2018zxxjqrjydybg.pdf.

Science Promotion Committee of the Chinese Institute of Electronics. 2019. *2019 Report on Robotics Education in Primary and Secondary Schools*. December. http://www.kpcb.org.cn/h-nd-343.html.

The State Council of the People's Republic of China. 2017. *Circular of the State Council on Printing and Issuing the Development Plan for the New Generation of Artificial Intelligence*. Accessed May 8, 2020. http://www.gov.cn/zhengce/content/2017-07/20/content_5211996.htm.

Wang, Rongliang. 2017. "Some Thoughts on the Implementation of Computational Thinking Education." *China Information Technology Education* 18: 21–25.

Zhang, Xiujie. 2013. "Research on the Instructional Design of the Robotics Education in Primary and Secondary Schools." Master's diss., Shenyang Normal University.

CONTRIBUTORS

Harold Abelson   Department of Electrical Engineering and Computer Science, Massachusetts Institute of Technology; hal@mit.edu

Michal Armoni   Department of Science Teaching, Weizmann Institute of Science, Israel; michal.armoni@weizmann.ac.il

Tim Bell   College of Engineering, University of Canterbury, New Zealand; tim.bell@canterbury.ac.nz

Shiau-Wei Chan   Faculty of Technology Management and Business, Universiti Tun Hussein Onn Malaysia, Malaysia; swchan@uthm.edu.my

Katrina Falkner   Faculty of Sciences, Engineering and Technology, University of Adelaide, Australia; katrina.falkner@adelaide.edu.au

Judith Gal-Ezer   Department of Mathematics and Computer Science, Open University of Israel, Israel; galezer@cs.openu.ac.il

David Harel   Department of Computer Science and Applied Mathematics, Weizmann Institute of Science, Israel; dharel@weizmann.ac.il

Ronghuai Huang   Smart Learning Institute, Beijing Normal University, China; huangrh@bnu.edu.cn

Megumi Iwata   University of Oulu, Finland; megumi.iwata@oulu.fi

Sridhar Iyer   Educational Technology, IIT Bombay, India; sri@iitb.ac.in

Ronak Jogeshwar   Atal Innovation Mission, NITI Aayog; ronakvj@gmail.com

Dongsim Kim   Hanshin University, South Korea; southpaw61@hs.ac.kr

Jussi Koivisto   Code School Finland; jussi@codeschool.fi

Siu-Cheung Kong   Artificial Intelligence and Digital Competency Education Centre, Education University of Hong Kong; sckong@eduhk.hk

Wai-Ying Kwok   Artificial Intelligence and Digital Competency Education Centre, Education University of Hong Kong; waiyingk@eduhk.hk

Jari Laru   University of Oulu, Finland; jari.laru@oulu.fi

Jia Li   National Institute of Education Science, Beijing, China; lijia515@163.com

Chee-Kit Looi   Department of Curriculum and Instruction, Education University of Hong Kong; cklooi@eduhk.hk

Kati Mäkitalo   University of Oulu, Finland; kati.makitalo@oulu.fi

Rami Marelly   Plethora Technologies Ltd., Israel; rami@iamplethora.com

Juan-Carlos Pérez-González   Universidad Nacional de Educación a Distancia (UNED), Spain; jcperez@edu.uned.es

Ashutosh Raina   Educational Technology, IIT Bombay, India; raina.ashu@iitb.ac.in

Youqun Ren   Institute of Curriculum and Instruction, East China Normal University, China; yqren@admin.ecnu.edu.cn

Marcos Román-González   Universidad Nacional de Educación a Distancia (UNED), Spain; mroman@edu.uned.es

Dahyeon Ryoo   Ewha Womans University, South Korea; waitemoon1@ewha.ac.kr

Jeremy Scott   Principal Teacher of Computing Science, George Heriot's School (retired on August 22, 2021), United Kingdom; jeremy@jtscomputing.com

Peter Seow   National Institute of Education, Nanyang Technological University, Singapore; peter.seow@nie.edu.sg

Ju-Ling Shih   Graduate Institute of Network Learning Technology, National Central University, Taiwan; juling@cl.ncu.edu.tw

Smadar Szekely   Department of Computer Science and Applied Mathematics, Weizmann Institute of Science, Israel; smadar.szekely@weizmann.ac.il

Hyo-Jeong So   Ewha Womans University, South Korea; hyojeongso@ewha.ac.kr

Matti Tedre   University of Eastern Finland, Finland; matti.tedre@uef.fi

Teemu Valtonen   University of Eastern Finland, Finland; teemu.valtonen@uef.fi

Rebecca Vivian   Faculty of Sciences, Engineering and Technology, University of Adelaide, Australia; rebecca.vivian@adelaide.edu.au

Bimlesh Wadhwa   School of Computing, National University of Singapore; bim lesh@nus.edu.sg

Su Wang   Institute of International and Comparative Education, National Institute of Education Science, Beijing, China; bjwangsu@126.com

Longkai Wu   Faculty of Artificial Intelligence in Education, Central China Normal University, China; longkaiwu@ccnu.edu.cn

Guangde Xiao   School of Education, Hebei University, China; xiaoguangde@163.com

Yudhisther Yadav   Strategic Initiatives, NASSCOM Future Skills!; yadav.yudhisther@ gmail.com

Junfeng Yang   School of Education, Hangzhou Normal University, China; yjf@hznu .edu.cn

Xiaozhe Yang   Institute of Curriculum and Instruction, East China Normal University, China; worldetyang@gmail.com

Hui Zhang   Faculty of Education, Beijing Sport University, China; huizhang@indiana .edu

INDEX